AF615118

ANY RESEMBLANCE TO ANTIBIOTICS' FATE IS A MERE COINCIDENCE

Plague

IMMUNE CROSSOVER

The Two Faces of Immunity

Dedicated to

Herbert Begemann (1919–1994)

Humane visionary and outsider

Everything in the universe interconnects, immunity being the exception.

Thoughts about collective interactive immunity: an interdisciplinary speculative perspective.

IMMUNE CROSSOVER

The Two Faces of Immunity

An Approach to the Dangers of Plague

Enrique Rewald

The Parthenon Publishing Group
International Publishers in Medicine, Science & Technology

NEW YORK LONDON

Library of Congress Cataloging-in-Publication Data

Rewald, E.

Immune crossover : the two faces of immunity : an approach to the dangers of plague/Enrique Rewald.

p. cm.

Includes bibliographical references.

ISBN 1-85070-018-4

1. Immune system—Evolution. 2. Ecology. 3. Immunological tolerance. I. Title.

[DNLM: 1. Immunity. 2. Ecology. 3. Communicable Diseases. QW 540 R454i 1998]

QR182.2.E94R48 1998

616.07'9—dc21

DNLM/DLC

for Library of Congress

97-43924

CIP

British Library Cataloguing in Publication Data

Rewald, Enrique

Immune crossover : the two faces of immunity : an approach to the dangers of plague

1. Immunology

I. Title

616'.079

ISBN 1-85070-018-4

Published in the USA by
The Parthenon Publishing Group Inc.
One Blue Hill Plaza, PO Box 1564
Pearl River
New York 10965, USA

Published in the UK and Europe by
The Parthenon Publishing Group Ltd.
Casterton Hall, Carnforth
Lancs. LA6 2LA, UK

Typeset by Blackpool Typesetting Services Ltd., UK
Printed and bound by Bookcraft (Bath) Ltd., Midsomer Norton, UK

Acknowledgements

Invaluable insight and expertise were provided by the distinguished immunohematologist Carlos Alberto González of the Francisco Javier Muñiz Hospital, from data gathered by the experienced immunohematological technicians Liliana Guzmán and Gabriela Nocetti. Urs Nydegger's contribution also deserves a special mention.

Thanks are due to my long-standing co-worker María de las Mercedes Francischetti, my distinguished colleague Fermín Enrique Domínguez, the Head of the Department of Physics at the Universidad Nacional de Mar del Plata, Alberto de la Torre, the respected staff from the library of the Centro Médico de Mar del Plata under the guidance of Isabel Bonzano, literature-teacher Susana Petrazzini, the late physicist José Kleiner (1910–1997)—a former pupil of Albert Einstein and Max Planck, my secretary Ana María Harvey, and especially also to the people from Parthenon Publishing. Furthermore, I offer a particular feeling of gratitude for Margarita, my wife.

Contents

INTRODUCTION

Figure 1 After such an eternity, do you think that my new outfit suits me?

Genes and 'Milieu'

Everything in life needs balance. The outcome of any extremist position, including one-sided scientific ones, is dubious at best; crowded living conditions are clearly such an extreme, and whenever mentioned people will be antagonistic to them. It is not inconceivable, however, that among all the drawbacks of overcrowding something positive could be found. This book will focus on thoughts about possible behavioral adaptations of immunity in the context of ecology[1], and how immunity may overcome conventional boundaries.

Recent insights in the new discipline of psychoneuroendoimmunology have led research efforts to concentrate on immune repercussions, or, to put it another way, mutual immune influences, through indirect means, notably via the psyche. In particular, stress is being examined from this point of view. There is also the possibility of 'placebo' links.

The fear that there is still a real possibility of a sudden global outbreak of plague was comprehensively documented by Laurie Garrett in her book *The Coming Plague* [2]. The present book focuses on certain immune defense capabilities in connection with the perceived risks; the situation is not, however, without hope—indeed, it might be appropriate to think in terms of the analogy of a glass of water being half-full instead of half-empty.

FIGURE 2 HALF FULL OR HALF EMPTY?

Mankind takes part in the food-chain like any other species and has paid a high price to evolve, the standards of nature being, in general, extremely wasteful. Think, for example, of the ratio of spermatozoa to fertilized ova. Nevertheless, overall, human life appears to have evolved in apparent equilibrium with the environment. To have thrived over the millennia in hostile surroundings is an astonishing achievement. The conversion from rural dwelling to the 'mega-city' also provides evidence of a superb adaptability; nevertheless, an ever-present sword of Damocles hangs over us.

Immune mechanisms are by no means invariant, and a response against a specific intruder varies between individuals. It depends, of course, on the genes encoded by the major histocompatibility complex (MHC) as a main determinator as to which epitopes are selected to react. In the course of life the immune system develops an astonishing flexibility in dealing with threats. But does this imply that it is totally confined within the body? Or is reciprocal interaction with other individuals a possibility? There is no doubt that inside a habitat in which humans are tightly packed, immune messengers from other hosts interact at the body surface and at orifices. The question is: can such contact influence our immune function? To explore this an interdisciplinary approach is essential, and the present work is an attempt to provide insight into this field of knowledge where academic boundaries blur.

Humans have always had to deal with disease. Recent genetic research has brought to light how germs learn, chameleon-like, to disguise their antigenic determinants. Usually, however, the host's defense mechanisms tend to keep pace. The disproportion in life-span between pathogen and host allows the former to mutate considerably more often[3], and accounts for changes in immune and flora constellations in the latter, which struggle to compensate for the handicap, at least partially, by the adaptability of their immune responses. One of the purposes of this book is to consider our long evolution in this respect, focusing on putative collective immune defense mechanisms, bolstered up, perhaps, by a common micro-flora—which could even include the livestock with which man has associated more recently in his evolution. Since everything in nature appears to be interconnected, a totally individualistic immune machinery sounds unlikely. Even microbes take advantage of the situation by forming colonies more often than was previously thought.

Similarly, a distinction should be made between people living in spacious surroundings and those forced to share insufficient space. Unhygienic overcrowding is associated with a multitude of adverse factors; however, it is possible that under such conditions invaders intermingle, in the process somehow stimulating the hosts' immune defenses. In the enforced closeness of crowded living, groups of invading pathogens—rather than mere individuals—may confront the mass of humanity together with its microflora which is often of long standing. Such an untested line of inquiry may not to some seem legitimate but this is the approach of the current work, which takes as its main leitmotif the concept that on many occasions defense is enhanced by the hosts acting in concert. Interspecies cooperation cannot be dismissed; one may in fact have to deal with a shared immune information–decision-making mechanism, and perhaps with a sophisticated collective immune strategy. By concentrating on the notion of a self-contained immune system[4], totally inside the body, perhaps too little attention has been paid to other approaches.

The idea that antibiotics have almost reached the end of their useful life is already popular; antibiotics are perceived as victims of their own success. For the moment, fears about them seem to be justified, though it is legitimate to ask whether most of them have already been discovered. Paleontologists discovered that evolution—of, for example, humanoid skulls or dinosaur snouts—seems to have proceeded at a leisurely pace, as a result of which it was never imagined that today's antibiotics and insecticides would so quickly become obsolete. However, genetic malleability increases with the number of microbes, their profligacy and their short life-span, and this may account for the fast acquisition of resistance to chemicals by today's aggressors. Furthermore, it has been ascertained that, at least in the fruit fly, resistance to cyclodiene appears to have originated before insecticides were introduced. Acquired molecular defense mechanisms can be passed on to others individuals, even those of different species. Most critically, drug resistance that is caused by a single point mutation can be quickly spread through modern transportation facilities. In classic Darwinian fashion, the more antibiotics and insecticides are used, the faster the defense strategy of microbes will evolve. Those that adapt will survive and proliferate.

During its overall life-cycle, an organism is exposed to a vast variety of pathogens in their diverse stages of development, mutation and evolution. For better or worse, the microflora communities living on body surfaces and orifices may expand and perhaps interact when people's habitats become congested. The 'milieu' may also attract germs from its surroundings. Microbes influence each other in different ways: they may repel each other, aggregate in clumps, etc. Interactive processes evolve in the habitat among species—particularly in the relatively stable local flora. Most of the species are optional saprophytes, but opportunism is important in species' ecology. Danger does not necessarily come from afar. Diversity, omnipresent in nature, also has its effect on long-standing flora components, and this diversity, along with their potential to spread, complicates matters.

It is an unfortunate fact that it is impossible to see the whole of reality at the same time, so investigators can only examine a part of the whole. Thus, in any field of research, in the absence of easily accessible evidence, it becomes necessary to speculate actively to promote investigation. So far, there are no proven circumstances in which external immune relationships are relevant, but it should be remembered that in the early days of conventional immune research, advances were made through speculation. Two opposing letters appeared in the London *Sunday Telegraph* of 10 August 1997. One letter claims that 'at last the truth about the "chemical soup" in we live is getting some publicity'; the other (acting as Devil's advocate) cited Bruce Ames[5], who has stated that 'roughly half the natural and synthetic chemicals ever tested are carcinogenic and, interestingly, 99.9 per cent of pesticides that humans ingest are natural.' In other words, the quantities of natural pesticides dwarf those of synthetics, and yet neither is poisonous in everyday life. Our bodies' defenses against chemicals are on the whole general rather than specific. Ames denies that synthetics have any special potential to cause harm, as environmentalists often claim, adducing evidence from America that shows that the quarter of the population eating the least fruit and vegetables (which are said to be laced with pesticide residue) has double the incidence of most types of cancer compared with the quarter eating the most fruit and vegetables.

From the subatomic up to the astronomical levels, nature reveals interdependence, or, rather, interconnection. Although in the past some gifted researchers seriously attempted to take a general approach, this has not been the norm, and the present quasi-systematic progression of knowledge requires joint efforts to bridge disciplines and to counter the narrowness of independent single-topic research.

The aim of this book is to document these concepts and to provoke our thoughts.

BACKGROUND

Immune Memes*

The ever-increasing body of data that relate to the immune profile has added to the complexity of the subject. The immune system evolves in parallel with the host as a whole and may develop further in order to effect any changes needed for its continuing adaptation. As a consequence, the task of working and communicating in this field is becoming a considerable challenge. When, in the 1960s, I collaborated with Ludwig Heilmeyer's and Karl-Georg von Boroviczény's initiative for consensus in hematology[6], I learned the

FIGURE 3 SELF-EVIDENCE IN TODAY'S IMMUNOLOGY

* Meme: 'an idea, behavior, or style usage that spreads from person to person or within a culture'. *Webster's Dictionary* (forthcoming)

importance of global understanding. Common terms with universal acceptance, when clearly defined, are of paramount importance.

A number of 'memes' pertinent to the text are briefly outlined here:

'Adjuvant': a substance which enhances a response.

'Antibody': a serum protein adapted to specific binding to the immunizing epitope.

'Antigen': a material with potential to induce a specific immune response.

'Antigen-presenting cell' (APC): a cell that expresses on its membrane class II major histocompatibility complex (MHC) molecules involved in processing and presentation of antigen-to-helper T-cells.

'B-cell': a cell destined to produce, to carry and to release immunoglobulin; it also expresses class II MHC on its surface.

'Chimera': reference to genetically distinct cells in an individual. The term derives from the myth of an animal possessing the head of a lion, the body of a goat and the tail of a snake.

'Class I, II and III molecules': proteins encoded by genes in the MHC.

'Complement': a number of serum proteins that activate in a cascade order that may be triggered either 'classically' by the interaction of an antibody with specific antigen or an 'alternative' (shorter) pathway through non-specific stimuli.

'Cytokines': cell-secreted messenger molecules acting in cells' 'talk'.

'Epitope': a single antigenic determinant.

'Idiotype': the antigenic determinant (idiotope) in the variable region of an antibody.

'Integrin': a member of a family of cell surface adhesion molecules that bind mainly to extracellular matrix.

'Interleukins': glycoproteins produced by diverse white blood cells that mainly react with other leukocytes.

'Lipopolysaccharides' (LPS): a component of Gram-negative bacteria that induce 'trigger-happy' host reactions.

'Major histocompatibility complex class I' molecules (MHC-I): take part in antigen presentation to cytotoxic T (CD8 +) cells.

'Major histocompatibility complex class II' molecules (MHC-II): take part in antigen presentation to helper T (CD4 +) cells.

'Naïve lymphocytes': lymphocytes that have developed but have yet to be activated by antigen.

'Pheromones': chemical messenger molecules that many species emit to communicate with other members of their own and other species, and also for self-orientation.

'Selectins': cell-surface adhesion molecules diversely binding carbohydrates.

'Superantigen': an immuno-stimulatory molecule produced by bacteria and viruses. The characteristics of binding to the MHC outside the antigen-binding cleft, and of stimulating T-cells in a T-cell receptor, provide potent immune effects.

'T-cells': lymphocytes which differentiate via the thymus.

'Tolerogens': antigens with the capacity to induce tolerance.

Something Related to Mathematics, Physics, Chaos and Meteorology

A unique characteristic of life is that it is an organized system capable of creating more order from less order; although this may appear to contradict the second law of thermodynamics, it is not, in fact, the case. The second law applies only to a closed system, so an open system can interact with the environment, exchanging energy in the process. In a sense, we feed more on entropy than on energy.

The first real attempt to understand how global behavior might differ from local behavior came from mathematicians' multidimensional topology. We tend to focus on small population clusters that are more susceptible to chance fluctuations. In population dynamics, aggressors and hosts alternately gain the upper hand: through the development by hosts of strong defenses, pathogen numbers may be forced to decline, while lowered stimulation by an aggressor may cause immune alertness to diminish[7]. The process makes for a collective homeostasis and flexible immune regulation—a possible mechanism to attenuate an individual's overextended or deficient response[8].

One basic fact is that if as residual so-called 'free space' in a crowd should not simply be dismissed as empty; it fully deserves a careful scrutiny that includes its immunological aspects. It is quite understandable that many of the notions described depend on recent insights into what traditionally were called exact sciences. These insights come from new work which started with shifting views from linear systems to networks. Physics, in particular, inspires one to broaden the scope to examine the area of influence of immune networks. The multiple messenger stimuli subject is based on Bertrand Russell's (1872–1970) premise that every end has a beginning; here it is supposed to be linked to the microhabitat.

Inertia is a passive property that does not modify a body's state of motion, except to oppose forces tending to change its direction or velocity. The inertia of a body depends on its mass and its moment of inertia.

Physics shows that order can arise—i.e. entropy can decrease—locally if there is a concomitant increase in the disorder of the environment. Brack summarizes the conditions that lead to 'magic' stability in the context of the periodic table of the elements and refers to atoms or molecules showing a tendency to aggregate into clusters—a condition that is also shown by tiny metal clusters, all of which points to some kind of ordering process and raises many questions for the physicist[9]. Furthermore, such metal clusters can operate in the chemical industry as very effective catalysts. It should not, perhaps, come as a surprise that in a crowded habitat the distribution of particles also shows a kind of structure that could influence the hosts' defenses—though this may be hard to visualize, or even appear counter intuitive. Could such ideas be related to the way in which people confined in foul prison cells confronted the host of adversities afflicting them better than expected, possibly by creating localized order among the surrounding chaos?

Friction is a dissipative process. In space, free of friction, periodic motion is seen in the orbits of heavenly bodies; on earth, virtually any regular oscillation can be likened to some analog of the pendulum. Friction tends to reduce chaotic behavior. The question arises whether it is feasible or not that friction, through concentrated particles' pollution, might contribute to the decoy effect for pathogens.

In mathematics, fractals (fractus = broken) are any of a class of complex geometric shapes that commonly exhibit the property of self-similarity. As distinct from conventional Euclidean geometry (square, circle, sphere, etc.), fractals are self-similar objects whose parts resemble the whole, giving rise to a new system of geometry with impact also on physical chemistry, physiology and fluid mechanics. There are, for example, stochastic (or random) tree barks and snowflakes. Another key characteristic of a fractal is a mathematical parameter called fractal dimension, which remains the same despite being magnified or whether the angle of view varies. Fractal dimension reveals precisely the nuances of the shape and complexity of a given non-Euclidean figure. Fractal geometry with its concepts of self-similarity and noninteger dimensionality is applied in statistical mechanics, notably when dealing with physical systems consisting of seemingly random features, e.g., fractal simulations to plot the distribution of galaxy clusters throughout the universe or to study problems related to fluid turbulence.

COMMUNICATIVITY

The astonishing creativity and freedom of thought of infants have been acknowledged for a long time. Recent evidence suggests that children of 3 years of age may be best qualified to solve certain complex problems in science and initiatives have been established to investigate children's potential. Among the concepts that are prone to confusion is noncommunicativity. In many cases it injects a spark of anarchy into a tame, well-behaved theory. Often it turns up at the scene of major upheavals in the scientific view. The noncommunicativity of actions has implications ranging from the trivial to the profound. Alexander the Great (356–323 BC) vented his legendary impatience in untying knots by slicing asunder the Gordian knot. Looking at systems by examining isolated molecules, and then integrating the results, has no future. Moreover, noncommunicativity obstructs a proper understanding of the possibilities suggested by considering collective immune behavior.

'MILIEU SOUP'

Substances reach hosts by a combination of dispersal and settlement that is determined by two kinds of processes which are termed vectorial and stochastic. Vectorial refers to directional dispersal by environmental motion, such as in wind and water currents. Stochastic processes are essentially random forces the operation of which does not allow a prediction to be made as to where particles will be carried and will settle.

The term 'milieu soup' has been coined to describe a densely packed group of people that may interact to influence the action of parasites as the hosts are squashed more closely together.

Chaos Theory Basics for Immunity and Disease Spread

Would you enjoy listening to a single-note concert (Figure 4)? By allowing all the instruments to interact, chaos may result. But it may be an enjoyable chaos (Figure 5).

Like clouds staging unpredictable spectacles, much has to be learned about the spread of disease. What was thought to be an obvious process is not so simple as might superficially appear. Because many epidemiologists and bacteriologists are biased toward studying microbes, work on other factors involved tends to be neglected. Chaos poses problems that defy customary science[10]. To

FIGURE 4 THE SINGLE-NOTE CONCERT

FIGURE 5 ENJOYABLE CHAOS

the enthusiast, it shares highlights with relativity and quantum mechanics. As a physicist once put it: 'relativity eliminated the Newtonian [1642–1727] illusion of absolute space and time, Max Planck's [1858–1947] quantum theory overcame the paradigm of a controllable measurement process, while Laplacian [1749–1827] deterministic predictability was abolished by chaos.' Such a revolutionary intuition could be applied to nature as a whole, questioning everything we perceive. Just a heresy? Perhaps not. By turning from the Establishment's natural tendency to assume it is on the right track, the chances appear favorable further to broaden the scope to ask how order arises (spontaneously?) out of chaos. Further, the random relationship between what goes into the system and what comes out points to interference. Instead of working, as hitherto, like mainstream physics that simply considers particles' properties whether they are tiny or huge, a new approach might provide more light from a different direction. Gradually, it may become apparent that standard assumptions may not fit the progress of an epidemic. Minor discrepancies may build up into significant consequences.

Examples of this are the activation of the complement and the clotting systems, which both depend on the assembly of enzymes activating the cascades. *In vitro,* the activating substances (for complement: antigen–antibody complexes, lipopolysaccharides, mannan-binding proteins; and for clotting systems: thromboplastin, tissue factor) may be introduced into solutions of components up to quite high concentrations, which are nevertheless below those needed to activate the system; any further increase in the concentration of the activator suddenly leads to a level capable of 'firing' the whole system.

FIGURE 6 OUR IMMUNE MECHANISMS SHOW DETERMINISTIC FEATURES IN SPITE OF THE HUGE NUMBERS OF FACTORS EXPONENTIALLY MULTIPLYING TO VARIABLES

On the other hand, awareness of biology's ability to reestablish equilibrium is essential. To scrutinize pathogens as they move around the observed hosts as expected, does not provide a whole understanding of the fate of millions of people. To discover harmony in omnipresent randomness and complexity you have to be able to identify obscure patterns transcending time and space. Why should we abandon the many trails rescued from the past and from almost unknown human communities still to be found around the globe?

The complexity of the transition from smooth flow to turbulence comes as a surprise when observed in a laboratory experiment, but it is even more marked in a sprawling 'mega-city' with multiple air, land and sea connections. In Lorenz's (1854–1946) system a minimal error proved catastrophic. Two nearly identical outbreaks of disease developed along totally different lines, which might suggest that long-range forecasting of epidemics may not be possible.

In the case of outbreaks of disease, statements of fact are, of course, required, but they do not always provide secure forecasts of future trends. In fact, the spread and regression of 'flu are as complicated as any atmospheric process. Both cases have periodic components, such as the well-established tendency for a 'flu epidemic to occur during the cold season, although it is not impossible that winter may be warmer than summer. The problem is that prognosis depends on unpredictable factors which seldom have much weight individually—but, like a bomb being detonated, a tiny incident can trigger off a disastrous plague. The hope not only of understanding the initial conditions, but of extrapolating from them to make predictions for the future, is easily foiled; superficially simple initial conditions may lead to epidemics through intricate interactions. Epidemiology is on the verge of a big change. Statistics is well-established, and is better than nothing—indeed, many lives have already been saved through the application of statistical analysis. However, to improve our knowledge of the spread of disease would require a vast network of sensors to provide all the necessary data. Furthermore, the computers upon which we now place such reliance can malfunction with disastrous consequences. The question is: can we rely on anything?

Chaos theory has a central role in current attempts at understanding. It postulates sensitive dependence on initial conditions that the well-known 'butterfly effect' symbolizes. Tiny errors and uncertainties, multiplying through chains of turbulence, can in theory lead to this phenomenon, rather like the changes engendered by giving an extra shuffle to a pack of cards before dealing them. The butterfly effect is no accident, it is to be expected and is rooted in ancient folklore:

> For want of a nail, the shoe was lost.
> For want of a shoe, the horse was lost.
> For want of a horse, the rider was lost.
> For want of a rider, the battle was lost.
> For want of a battle, the kingdom was lost.

For a long time it has been taken for granted that any whole made up of many smaller elements may change. When a chain of events is triggered by a moment of crisis, sensitive dependence on initial conditions becomes significant—it is an inescapable consequence of the way small-scale factors interact with large-scale factors. That would not be the case if the immune network could be reduced to a few understandable, deterministic laws ensuring predictable long-term behavior. Inasmuch as immunity depends on many distinct ingredients and irregular external influences, it will remain as an unstable nonlinear system. Nonlinear relationships are never simple to understand, while linear equations are both solvable and relatively simple to understand. In the immune network—with lots of known and lots of guessed factors—it is clear that divergent signaling by messengers may induce capricious 'out-of-control' interactions.

Lorenz found unpredictability, but he also found structure amid seemingly random patterns of behavior. An underlying order, therefore, has to be suspected. The disordered behavior may act as a creative process. By generating complexity, rigidly organized patterns convert to unstable, finite patterns to infinite. One could say that chaos has become not just a theory but also a method.

Through the work of many research centers, an overall chaos theory is emerging. Even the simplest things require re-evaluation, and, by considering the whole, you might discover a kind of order that Chaos specialists call 'determinism', be it waves in the ocean or music. This may influence the growing tendency towards specialization. To a certain extent chaos opposes believers in reductionism, who find the truth in the particle itself. Chaos theorists started to focus on star clustering, the behavior of the economy, micro-organisms, atoms, and so on. To decipher the particularities of their structure dynamics they use computers, graphics and other means.

For Aristotle (384–322 BC), physical motion was not a quantity or a force, but a kind of 'change'. Modern physics stresses that many kinds of observations only become clear if subliminal effects are considered. So the idea that confinement could facilitate immune contacts has to be thought of in terms of point interactions rather than in terms of integration.

A WORD FROM THE METEOROLOGIST

An interdisciplinary approach, nevertheless, has no obligation to include everything. Biometeorology and meteoropathology have been observed from ancient times and mentions can be cited from the Old Testament and authors in antiquity. Although they provide a different view of the remarkable tendency of living things to maintain homeostasis, it comes as a surprise to find that climate does not exert a major influence on the progress of epidemics. It is also a fact that people are divided almost equally between those who are very sensitive to minimal microphysicochemical and macrophysicochemical environmental changes and those who are apparently invulnerable to significant atmospheric variations, geomagnetic storms, etc.

A Brief Review of Historical Thoughts Related to Interactions

SOUL - MIND - BODY

Plato (422–347 BC) believed that the soul was fundamentally distinct from the body, although often it was affected by its association with the body. In spite of being simple, rather than composite, and thus not liable to dissolution, as were material things, the soul had the potential to rule the body. Plato denied the Pythagorean (570–497) view of the soul as an 'attunement' of the body. For Aristotle the soul was neither material nor immaterial but an abstraction. In plants this was concerned solely with nutrition and reproduction, but in animals were added sensation and independent movement, while in men the soul's crowning achievement was rational activity, although this was also thought to be under outside influence.

From René Descartes (1596–1650) onward more was heard of the mind than of the soul—the individual's mind especially. Descartes held the view that mind and body are capable of affecting each other causally, so that what happens in the body can produce effects in the mind, and vice versa. He stated his belief that one's soul is not to be found in the body like the captain is in a ship, but was more intimately bound up with it. We may remember the famous *cogito* argument 'his only certainty is the existence of his mind, for the very act of doubting was itself mental'. The Cartesian account of mind and body had many critics even in Descartes' own day when, for example, Thomas Hobbes (1588–1679) argued that nothing existed except matter in motion[11,12].

Baruch Spinoza posited a single entity, God or Nature, possessed of infinite attributes, of which the mental and the material alone are only known to men[13]. Its manifestations were the result solely of its nature, and arbitrary will neither did nor could play any role in them. Whatever manifested itself under one attribute had its counterpart in all others. According to David Hume (1711–1776), minds and bodies alike were nothing but 'bundles of perceptions', interactions between which were always possible. Starting with Georg Wilhelm Friedrich Hegel (1770–1831), accounts of action and cognition became less dogmatic, with a closer inspection of the facts.

John Locke (1632–1704) thought that 'sensible' ideas of size, shape, position and motion or rest resemble material objects as they are, and this is in agreement with modern science which is based on the mathematical focus on primary properties. Locke's ideas were further evolved by Hume, who put mathematical objectives above those founded on emotions. Hume's distinction between fact and value issued from the elevation of the mathematical objective over the emotional and the subjective.

At the beginning of the twentieth century the belief that mind and matter are constructed out of neutral monads still reveals the influence of the early Cartesians. At the present time machines have also become involved. They do not, of course, literally think, but they certainly come to conclusions. In the light of their 'experiences' they can improve their performance. They even have an analog of consciousness in the sensitivity they show to external stimuli.

Like any biological process, immunity and its dynamics are subject to modification—not only in the context of determinate milieu, but also with time. We know very little about the entire immunological orchestration in historical terms, but it would be naïve to deny totally the possibility of modifications. We must keep alert because the characteristics of plague are subject to dynamic change as well, and any search for a strict parallelism would be handicapped. Although the cognitive achievements of immunity are impressive, it is not good to attribute its deeds mechanically.

SELF

Stress on the 'self' or 'ego' is behind the developmental idealism of Hegel, who envisioned a World Soul coming to consciousness. John Paul Sartre (1905–1980) argued that each individual chooses to come into being out of nothingness. He upheld the Cartesian position that the self is conscious by denying the unconscious self proposed by Sigmund Freud (1856–1939). For Gilbert Ryle (1900–1976) the mind (ghost) is simply the intelligent behavior of the body.

INTERACTION

Interaction is a universal phenomenon—for example, even at the cosmological level certain pairs of galaxies interact (with spectacular results for astronomers). Although interaction is deeply rooted in biology, retrieval may be difficult. One implication of interactionism is that there cannot be a complete explanation of population defense mechanisms exclusively in terms of the individual's immune system. A broader view of immunity appears to support Levin and Solomon, who suggested that the body is simultaneously an evolutionary biological entity and an ongoing achievement of socialization in which dynamic intercommunication gives rise to either a healthy or a diseased state[14]. Medical thinking has focused sequentially on anatomical structure, physiological function and biochemical analysis, but now there is a trend to favor a more holistic perspective[15].

BIOTIC INTERACTIONS

No species lives in isolation from other species. The universal fight for survival, that includes the whole food chain, ensures a constant struggle; that plants and animals are attacked by parasites is part of it. Biotic interactions may occur between members of the same species (intra-specific) or between two or more species (inter-specific). Different species generally interact with each other competitively, but sometimes to their mutual benefit (symbiosis). Besides positive or negative effects upon the participants, the members of an interaction can remain apparently unaffected. Some interactions are temporary, casual and of minor importance, whereas others are permanent, vital and of major significance. The diverse interactions, which evidently include defense mechanisms, are often highly specialized and complicated, involving adaptive changes in structure, function, behavior and ecology[16]. Biotic interactions are significant not only because they influence individual species, but also because they constitute the principal stabilizing, connective linkages among the various species contained in a community. In a general sense, species in a biological community maintain a relative harmony because of biotic interactions, despite individual gains and losses. As a consequence, the community as a whole persists, with all the individuals contributing in some way.

Mutualism is an association of two species that results in mutual benefit or gain. Facultative mutualism, or protocooperation, refers to an association in which one member can survive in the absence of the other. Commensalism means that one member benefits while the other remains unaffected.

Early Evolution of Host Defenses

The immune system is inseparable from an individual's integrity and therefore takes its phylogenetical roots from the evolution of most primitive organisms. Single cell organisms defending themselves against the environment—although they do not have a distinct cell organelle which could be ascribed to the immune system—behave immunocompetently insofar as they protect their integrity against the outside world. Higher plants and lower animals possess tissue antigens that prevent simple transplantation of parts, even when derived from individuals of the same species. During the course of their evolution, animals develop specialized defense systems, and it is well known that marine invertebrates possess proteins of the alternative pathway of complement[17], and this in the absence of any specific immune response. Higher up the evolutionary ladder, defense mechanisms increase morphological and functional efficiency. The systemic patterns they deploy are called immune systems.

The paramount importance of chemical communication in biological systems is now widely appreciated. Messenger molecules reach their target receptors *via* body fluids and, although to a much lesser degree, perhaps by circumventing external milieu. Awareness of diversity is still on the rise. Many cell types share receptors of diverse affinity and many receptors may fit a specified molecule.

Pheromones are messenger substances secreted to the outside world by an individual, prompting a specific reaction in the recipient; in particular, they are well known in the life of social insects. Certain pheromones may have their equivalent in immunity, for example, those for alarm, or perhaps, straightforward defense. It would be exciting, if, in immune relations, an equivalent for tagging of aggressors by pheromones, seen in social insects, could be identified. Such tagging represents the means to alert other members of the community.

Given that all living creatures are made up of basically similar building-blocks, the omnipresent capacity to distinguish 'self' from 'not-self' is impressive. It should also be considered how remarkable it is that many active molecules from all kind of species work in similar ways. Both the individual and shared features of immune response are shown to be species efficient. It could be that such evolutionary conservatism has practical implications on the environment's immune properties. Interconnection is essential, and chemical signals in their widest sense are a universal attribute of life within and between all organisms. The great differences in the capacities of animals to analyze foreign material are reflected in the diversity of gross structural organization involved in defense mechanisms. In spite of that, their evolution provides remarkably equivalent capabilities to all species from the unicellular upward. Clearly, common defense features have vital implications in evolutionary conservation.

In fact, long before there was an awareness of the immune network, people realized by common sense that infection may spread through contagion and that in a number of diseases, when recovery is achieved, it means long-lasting protection. From early times this observation led to the development of biological warfare by catapulting the bodies of casualties of pestilence into besieged cities. Thucydides (*c.*460–400 BC) narrates how during the plague of Athens in 430 BC survivors attempted to provide help to the affected masses[18]. But identification of immune mechanisms awaited the dawn of modern biology. It was only in the late 19th century that linear antigen–antibody interactions were conceived. Shortly afterwards Metchnikoff (1845–1916) disclosed the importance of phagocytosis. Through observation of a rose-thorn penetrating a transparent starfish larva he became convinced that the ameboid cells attracted to the site of injury had evolved specifically to defend the starfish's integrity.

This chapter should allow the reader to understand the need for a specific cell system that is designed for defense. Sessile plants and animals, incapable of fleeing from predators, need on-site defense, which is provided by primordial immune reactions. It is on the basis of this observation that the evolution of immunity has to be seen, that is, mobile creatures retained these primitive immune mechanisms and developed them into sophisticated systems.

FIGURE 7 PHAGOCYTOSIS EVIDENCED BY STARFISH TRANSPARENCY PERPLEXES METCHNIKOFF

A landmark in the evolutionary process of immunology is coincident with the appearance of vertebrates. What is called the 'innate system' became supplemented by the 'adaptive system' that retains specific memory for stimulatory determinants. A feature of 'innate recognition' is the ability to bind to the components that are common to different microorganisms. The innate system always responds in the same way and time, irrespective of the number of previous exposures.

The ideas of immune recognition peaked by conceiving a foreign antigen-dominated idiotypic–anti-idiotypic network[19]. Step by step, different cell types and activating/suppressing molecules were incorporated into the immune response and links with inflammation, disease and other organ systems became evident.

There is no doubt that infection is far more noticeable when expressing disease than when aborted thanks to a successful immune reaction.

SOMETHING BASIC ABOUT IMMUNITY

Evolution has created a dispersed system of highly specialized immune environments, interconnected through an elaborate system of cellular targeting, molecular interplay and recirculation. The notion of strict adjustment to conventional anatomic limits is questionable at least. Much depends on the chemical structure and also molecular shape and size. It is now apparent that the host is endowed with a complex variety of adhesion receptors and chemoattractants that direct various leukocyte subpopulations into diverse tissues. Selectins initiate cell attachment along vascular endothelium, by mediating leukocytes traversing and rolling along inflamed endothelium, and CD11/CD18 integrins achieve firm adhesion to vascular wall. L-, P- and E-selectins react with sialyl Lewis x (sLex) but only E- and P-selectins react with its isomer sLea. Complex interactions at these levels have already provided a lead to develop novel cell adhesion inhibitors by homology modeling. A complex apparatus of selectins, integrins, sialomucines and, last but not least, the immunoglobulin superfamily provides for a finely tunable redistribution of immune cells both into inflamed tissue segments and/or throughout the whole host.

Recent insights into certain molecular steps are relevant: for example, interactions with C3b-receptors on cell membranes leading to signal transduction of activated transcription in the cells, with the intervention of tyrosine kinase[20]. The physiological network is revealed by immune cells, activated by immunogenic stimuli, that produce neuropeptides, ACTH and gamma-endorphin[21]. All the processes require an appropriate framework: for example, overstimulation can have noxious effects.

The integration and control of systemic immune responses depend on the regulated trafficking of lymphocytes. This ‘homing’ process disperses the immunologic response and directs subsets to the specialized microenvironments that control their differentiation and regulate their survival, as well as targeting immune effector cells to invasion sites. This exquisite targeting is determined by a combinatorial ‘decision-making’ process involving a multi-step sequential engagement of adhesion and signaling receptors. These homing-related interactions are seamlessly integrated into the overall interaction of the lymphocyte with its environment and participate directly in the control of lymphocyte function, life span and population dynamics.

Thus, the immune network appears in fact to be generalized and may even extend beyond conventional body limits, reflecting the daunting challenges that any immune system faces in its mission. A role for the immediate milieu—for example, the exact characteristics of the spread of disease, development of resistance or the adaptive behavior during the course of an illness—remains elusive[22].

Antigen Presentation and Lymphocytes

Individuals have their own identity. This polymorphism mainly evolved through natural selection, with mutation also contributing. Although evidence in outbred populations is hard to establish, by offering an even greater variety of epitopes, microbes with huge numbers of antigens should exert selective pressure on MHC polymorphism.

Subjects differ in disease susceptibility and disease resistance both in general and in particular to an infectious disease. In the laboratory, T-cell proliferation assays reveal high and low responders to an antigen. By crossing both variants Mendelian inheritance patterns come to light. Heterozygocity (maternal and paternal alleles of MHC molecules being codominant) assures a great variety. Grafts between allogeneic subjects trigger immune responses by virtue of cell-surface antigens, called histocompatibility antigens. Different genetic loci are involved in an individual's specificity. The molecules that induce the most rapid rejection of grafts are called MHC, so named after genetic mapping that revealed a cluster of several closely linked genes. The fact that the genes that control histocompatibility and immune responsiveness are identical is important. The MHC molecules belong to the most polymorphic molecules known. The main ones, class I, HLA (human leukocyte antigens) and class II, although only encoding about 10 per cent of the molecules, are the classical ones. Many more genes encode molecules of the same family with diverse functions, many of them still unknown. MHC functions are highly specific. Therefore, MHC genes are under very tight and complex regulation.

The relative abundance of an allele in a population seems to reflect its role in protecting against parasites found in its habitat. Among multiple possible mechanisms, for example, HLA B 5301 is most likely to present a certain plasmodium-derived peptide inside parasitized hepatocytes. By identifying the infected liver cells, these can be killed by CD8+ cytotoxic T-cells. An example of Darwinian natural selection provides several class I and class II HLA alleles that seem to have a role in the protection against *Plasmodium falciparum*. By killing people with normal red blood cells, the parasites allowed the African population carrying the sickle cell trait to increase substantially—similar to people affected with the diverse thalassemia, mainly concentrated in the Mediterranean Basin.

Class I molecules are present on almost every cell type, exacerbating under the influence of gamma interferon and other inflammatory cytokines. Normally, class II molecules are only expressed on dendritic cells, the prototype of antigen-presenting cells, macrophages/monocytes and B lymphocytes. In the thymus they are constitutively expressed by epithelial cells. Peripherally, inflammatory sites expand their expression to cells of epithelial origin. Potentially, any cell is suitable for class II activation, either by viruses or by mutations. There is no doubt about the importance of CD8 T-cells in their role in removing undesired cells (proper or heterogenous), and perhaps in removing detritus.

First, antigens have to be processed to peptides by antigen-presenting cells, reacting to the T-cells by proliferating (provided that the antigen-presenting MHC molecule on, for example, a dendritic cell (DC) belongs to the same specificity). MHC restriction means that T-cells only will react with peptides that present self-MHC molecules. In their tridimensional structure, class I and class II molecules have a singular similarity that is suited to peptide presentation. To express critical membrane receptor molecules, the maturing T-cells rearrange their antigen receptor genes while developing in the thymus. On release, the matured T-cells bear a receptor with affinity for the MHC. In order to respond, their dependence on MHC specificity obliges them to recognize the presented epitope jointly with the MHC. Perhaps the most significant distinction between class I and class II molecules lies in the phenotype of T-cells to which they present the antigenic peptide. CD4 T-lymphocytes are restricted by class II molecules, and CD8 T-lymphocytes are restricted by class I molecules. Cytotoxic CD8

T-cells only kill targets expressing antigen with syngeneic class I molecules. Similarly, CD4 helper cells (T_H) will only proliferate when the antigen-presenting cells have syngeneic class II molecules. A T-cell receptor, to react with an appropriate peptide–MHC complex, induces the CD4 or CD8 molecules to move in the plane of the plasma membrane to join the antigen receptor. Thus CD4 and CD8 are important accessory molecules for the T-lymphocytes' antigen receptors (they are also known as 'co-receptors'). Furthermore, enzymes involved in 'lymphocyte activation' associate either with CD4 or CD8 in the respective cytoplasmic domains.

To act, CD8 T-cells (specific cytotoxic lymphocytes), require awareness of autologous class I molecules in conjunction with the antigenic determinant of the target to be destroyed. To exert 'surveillance', that is, elimination of unwanted body cells, their antigenic stimulus must act in conjunction with class I molecules from the individual. That appears to be the 'reason' why the body of the entire cells is equipped with class I molecules. We may call them 'identification molecules'. Mostly they abound on immune cells; interferon-γ and other inflammatory cytokines enhancing their expression. On the other hand, class I molecules are scarce on other differentiated cells—for example, in the brain.

When MHC class II genes are absent or any of their developmental steps fail from regulatory promoter genes onwards, a deficiency or 'bare lymphocyte syndrome' results, that is to say an inability to generate T-cell-dependent T-cellular and humoral immune responses. In these patients, viral, bacterial or fungal infections already manifest in the first year of life, most probably involving the respiratory and gastrointestinal tracts.

Clonal expansion of CD4 T-cells may favor evolution towards T_H-1 cell type (cell-mediated orientation) or T_H-2 cell type (humeral orientation). These subtypes are mainly characterized by differences in cytokine production. Usually the branches T_H-1 and T_H-2 tend to be in balance, representing the extremes in cytokine release. T_H-1 preferentially produces IL-2, IFN-γ, TNF-β, GM-CSF and IL-3, helping cytotoxic T-lymphocyte proliferation, natural killer (NK) cells activation and promotes delayed-type hypersensitivity (involved, for example, in the inflammatory response of the skin). T_H-2, besides being the source of GM-CSF and IL-3, secrete IL-4, IL-5, IL-10 and IL-13, thereby helping antibody production (B lymphocytes), eosinophil production and mast-cell degranulation (see atopy below). Furthermore, there are intermediate cytokine-producing CD4 + cells which may be called T_H-0. Although the classification into T_H-1, T_H-2 and T_H-0, mainly depending on their cytokine-releasing pattern, is somewhat arbitrary, once a cell has established its cytokine profile it remains fixed and it cannot change to another profile type.

TOLERANCE

Immune responsiveness manifests diversely: it may evolve in phases or vanish. Immunologic tolerance regarding one or more specific antigenic determinants is manifested by diminished expression of either cell-mediated or humoral immunity[23]. To choose a certain antigen to mount an immune response may also depend on accessory molecules through adherence, or by delivering direct co-stimulatory signals. The latter may be provided by the network from the host, from subjects nearby or even come from other species, including microflora. Although it is not yet clear why one antigen may be ignored and others induce tolerance, tolerance of the non-dominant antigens is a side effect of antigen selection and serves to avoid autoreactivity. In early life, contact of self antigen with antigen receptors on immature B- or T-lymphocytes induces death or long-lasting unresponsiveness of these cells and thus clonal deletion* or anergy. Conversely, it is likely that peripheral tolerance can result from multiple molecular interactions, to which antigens (although not necessarily antigens) and many diverse additional signals contribute. Furthermore, a state of unresponsiveness termed anergy provides for further tolerance brought about by the absence of a secondary signal provided by helper T-cells or by the antigen-presenting cell[24].

* Clonal deletion is the loss of lymphocytes of a particular specificity due to contact with either 'self' or an artificially introduced antigen. From clone: (Greek *klōn* = young shoot or cutting): genetically identical cells or organisms derived by vegetative reproduction from a single parent; also, a DNA population derived from a single hybrid DNA molecule (recombinant vector) by replication in a eukaryotic or bacterial host cell.

MORE ABOUT IMMUNE INTERACTIONS

The ubiquity of the immune network is now widely accepted. By a kind of default, messenger immune molecules, hormones and even cells reach their targets in the tissues *via* body fluids. The diversity of the protagonists is still increasing; to a greater or lesser extent receptors are shared by many cell types. Factors relating to density, target cells and specific immune function may raise or depress immune responsiveness; they may also combine immuno-enhancing and immuno-suppressive effects. Signal transduction pathways have been observed, mainly inside tissue and body fluids[25].

All multicellular organisms have highly effective signaling systems enabling cells to communicate. The properties of networks emerge from their organization rather than from the characteristics of their individual components. The recent emergence of disciplines like psychoneuroendoimmunology (different names are given to this discipline) reveals an increasing effort through rigorous research to pinpoint how developed organisms orchestrate harmony and cooperation between the immune system and other important bodily functions. Neuromediators, hormones, immune messengers and cells gain access to many tissues by interacting with receptors that are by no means specific or confined to a certain cell type.

From a large but finite number of antigen–receptor-defined lymphocyte clones, the immune system must establish and maintain diverse, non-autoimmune populations of mature lymphocytes and endow them with the capability of antigen response. It must control the interplay between B-cells, T-cells and specialized accessory cell populations to orchestrate the response to antigens, as well as superantigens, so as efficiently to initiate primary cellular and humoral responses. It must integrate responses throughout the body, simultaneously facilitating specialization and regional modalities (for example, alimentary tract, lung, skin). Homeostasis provides long-lasting but malleable immunity over time, preventing overexpansion or depletion of specialized lymphocyte sets. For healthcare, an overview of effector sites and immune traffic seems of interest.

Nuances in the different elements deserve appraisal. For host-versus-germ reactions, it is not only the quality and spatial location of the epitopes that matter. Often antigens convert from basic form to whatever the milieu is able to modify them into, tailoring immune responses by different means. This could be another aspect to consider among possible causes that prevent some sex workers from contracting human immunodeficiency virus infection (HIV)[26]. Immune responsiveness against a specific virus varies between individuals and depends, among other things, on the genes encoded by the MHC that determine which epitopes of the virus are presented to the immune system. Immune responsiveness can modify over time, for example, in HIV as a result of antigenic mutation* or declining T-cells.

ATOPY†

In atopy, antigen presentation to T-cells represents an early stage in sensitization leading to selection of T_H-2-like clones, to selective production of specific IgE by B lymphocytes, and to eosinophil-rich inflammation in certain 'target' organs. Consequently, local physiological and environmental factors interact with these immunological processes to cause organ-specific symptoms. Apart possibly from primary sensitization, no step in the sequence is all-or-none. We will see later that immune features in the individual vary over time. Viral and bacterial antigens largely potentiate T_H-1 mechanisms, whereas atopic sensitization is effected by T_H-2-dependent mechanisms. Critical to our understanding of the potential interaction of infection and atopic sensitization is the current concept of the negative interaction between T_H-1 and T_H-2 subsets, although variations in an individual over time suggest that this might be an oversimplification. Much research has been done to determine if infectious diseases or their equivalent vaccinations can prevent atopic or allergic expansion.

BOUNDARIES

In spite of playing down the role of absolute boundaries, there is no question that conventional limits provide a first protection against

* Mutation is a change in the chemistry of a gene that is perpetuated in any subsequent divisions of the cell in which it occurs. Somatic mutation is a mutation occurring in the general body cells—rather than to the germ cells—and is not transmitted to progeny.

† A genetically determined state of hypersensitivity to environmental allergens. Type I allergic reaction is associated with the IgE antibody and a group of diseases such as hay fever, asthma and atopic dermatitis.

invasion. Skin, including the keratin-producing outer layer, is essential. Although less tough, respiratory and gastrointestinal mucosae also provide mechanical barriers. In addition, the mucosa-associated lymphoid tissue provides for a close examination of antigen invaders with the consequence, when a foreign type is identified, of reacting against it in a specific fashion. Glands cooperate, secreting oleic acid or lysozymes that may destroy bacteria. All the cell linings are constantly being renewed from underlying layers. Most endothelia have cilia to expel locally secreted mucus that may contain potentially harmful particles. Importantly, IgA immunoglobulin concentrates in mucus; it is at this level that the decision is taken to get rid of an antigenic substance, the host using its innate or acquired immunity or both. The preexisting mannose, lipopolysaccharide and scavenger receptors of mammalian macrophages might suffice to identify and defend against such invaders as yeast and bacterial cell walls. In addition, plasma proteins, such as those of the complement cascade, C-reactive protein and the mannose-binding protein of the collectin family also recognize microbial carbohydrates and defend against bacterial insult. In contrast to these preexisting and evolutionary structures, specific antibodies are produced through acquired immunity based on somatic gene rearrangement.

PROPOSAL

Immune Paradigms

FIGURE 8 SELF-CENTERED PERCEPTION OUTMODED?

Figure 8 depicts the human-centered view as a natural tendency, in our case, specifically pointing to immunity. Dogs mark their territory through particular pheromones that they spray when urinating. Chemistry, therefore, appears to be involved, but this is merely the starting-point for further studies. There is more—cognitive scientist Douglas Hofstadter has commented on a computer program that performs Mozart's '42nd Symphony' (Mozart only wrote forty-one): 'Technique has no model whatsoever of life experience, has no sense of itself, has never heard a note of music, has no trace in it of where I think music comes from. I am comparing that with an entire human soul, as forged by the struggles and travails of life, and all the experiences that create emotion'[27].

In the succeeding pages, immunology adheres to what we call a 'complex order'. Identification of new pieces of our 'immune project' led immunity to be centered in a positive image of 'self'. By this alternate paradigm, 'self' is permanently monitored through a compound network of natural autoantibodies and an increasing number of interacting factors. Actually, identification is possible, thanks to a series of major autoantigens that specifically characterize the diverse organs and tissues. In addition to their astonishing ability to detect deviations, they are now known to have a formidable regulatory mechanism[28]. These insights have deepened (but by no means comprehensively) our understanding of immune defense mechanisms[29].

When the Establishment switches to a new paradigm, it often happens that novel concepts are given undue importance and it usually takes some time for a proper balance to become established. After the emergence of the bourgeoisie in the course of the last century, the rising hierachy of people's identity became further emphasized in the 1920s by psychology. Gradually, post-Freudian thoughts evolved to a much broader approach which attributes a major role to community. Likewise, it may be felt that by shifting the paradigm toward the immunological homunculus, a rather similar overestimation of its importance occurred in immunology.

We must consider the individual as a particular aspect of community, like time and presence which were paramount in ancient Greek thinking and tradition. Looking at the individual in the context of the community closes the hermeneutic circuit. It follows also that in immunology a particular case has to be referred to the whole. Simultaneously, gene-dependent morphogenesis points to autonomous determinism that in turn suggests independence from external factors at first sight, supposedly incapable of directing and imposing common organization. Here we examine the impact of external influences, as opposed to individual genetic effects. The impressive number of environmental variants tends to conceal the identification of minor community-wide immune convergence.

Immune Cognition – the Immune Bias

FIGURE 9 PARIS' CHOICE FOR BEAUTY WAS APHRODITE. WHAT ABOUT IMMUNITY, DO YOU HAVE THE RIGHT INSTINCTS TOO?

Before starting, the following points may illustrate the aims of this section:

- until the silicon chip became available, carbon held the key for harnessing information (information meme);
- weak solutions may attract particles while high concentrations repel (chemical-recognition behavior);
- long-standing habitat sharing may induce synchronization of menstrual rhythms (erasing differences between individuals);

- some species actually smell water, while others depend on true hygro-receptors (unlooked-for perception).

A common-sense approach might suggest that it is quite reasonable that immunity strictly filters its choice of whether to focus on, or to ignore, a given signal. Effective operation of the system requires that large numbers of antigens, that also may present themselves in a camouflaged state, have to be disregarded; only a few, significant in their context, will be selected for treatment. Immunity does not just blunder into antigens, but, rather, orients towards suitable targets. Accordingly, the immune sense may become adapted to routine exposure of immune-signal emitters.

ABOUT INSTINCTS (OR PERHAPS A TELEOLOGICAL IMMUNOLOGY)

In immune defense, individuals' responses are often badly adapted to each other. Immunity is under the influence of many factors, often surpassing stimulation by naked antigens, *per se*. As with the conventional senses, so the immune network has to adapt to a routine. For example, when one looks at something, one does not become indiscriminately aware of most of the details in the field of vision; nevertheless, subliminally, many stimuli remain stored. A crowded habitat means steady exposure to immune messages and a complex routine of exchanging hosts' molecules and antigens. At length it may train insiders to challenge essentials and to build appropriate responses. Such a customary exchange is unlikely to remain without consequences. At present, it is not known if the hitherto-overlooked 'noise' exerts any influence when an organism chooses to immune-react. One interesting aspect of the cognitive capability of immunity is that, when responding, in fact it is usually to a combination of different stimuli. Often, related immune-recognition chains lead to similar immune behaviors, and hence, to variants of the same response. There is no doubt that cognitivity enhances the ability to survive hostile environments. Resistance is no simple consequence of exposure, as it mainly depends on complex cognitive decisions as an integral part of the defense system. It must be borne in mind that our immune system is under the reciprocal influence of other physiological networks. By extracting (intelligently?) information out of raw input and shaping experience, the capabilities of the immune system are shown to be very flexible and may even allow a reversal according to the circumstances. For example, a major concession to the 'self–non-self' principle is made in tolerating the implant of an heterologous embryo.

Immune subjectivity is not necessarily limited to the individual. When inter-host molecular traffic becomes critical, repercussions on immunological autonomy have to be expected.

It is worth saying that besides being able to show tolerance, acquired immunity can also become misdirected—for example, by attacking 'self'. Similarly, reacting against harmless materials such as food, pollen or dust is not uncommon. It is obvious that to be effective not only is recognition necessary, but so also is an appropriate response.

Although many data have been collected about qualitative and quantitative particularities in understanding antigenic signals, it is still not always clear what drives the immune system to choose between straight dismissal of antigens, refractoriness, or to attack furiously. The rules of tolerance induction clearly require further research. It is becoming increasingly clear that in immune decision-making, additional signals are involved which may not necessarily be directly related to the antigen. Perhaps individuals have the capability of sharing a common learning process.

Mutualistic immune influence through chemical and other clues

Focusing on cognitive properties implies the compounding of individuals with external factors. In a broad sense, externalism is more than the exchanging of messages, it even might serve to protect the weakest members from the group as a whole. Connection allows us to acquire knowledge and to learn how to behave; it also favors fostering. We suspect a role for tuning immune defenses. It is obvious that language does not need you, but can you imagine existing without it? Even bacteria reveal that some aspects of life are not restricted to the individual—in fact, most of the molecules that compound a micro-organism relate to communication, encompassing cognition and connection. As has already been said, cognition is central to immunity. From very low down the evolutionary scale[30], cognition starts with self-recognition, which is vital for defense. In every case the first question for a being is 'to react or not to react',

FIGURE 10 RIGHT AND WRONG

which is a basic principle in nature. Of course, signaling systems inside multicellular organisms with tissue specialization are far more sophisticated—'talk' among cells in the network context is very dynamic. If information can be conveyed, development of special codes, for example, by repetition, and ways of indicating readiness for external immune co-ordination, may exist. Our concern in this respect is the reciprocal immune bias that evolves in crowded conditions through permanent re-exposure to, and exchange of, molecular signals. As stated, immune decisions are inherently subject to error and, clearly, misinformation is a major handicap.

Here we can refer to Stuart Sutherland's book about human irrationality, *Irrationality: The Enemy Within*[31] which relativizes irrationality. It is an intriguing idea to translate some of his thoughts into the field of immunity. In addition to mistakes, like autoimmune disease or particular response dimensions (excess or deficit), the whole mechanical project of the system should be reassessed. Given the human tendency of Man to consider himself the center of everything, it appears that there is a remarkable parallel between the intellectual life and the immune life. By putting in doubt Man as a rational being we also have to consider that dimension of the individual concerned with the immune network image in relation to space, time, etc. The question is: by closing ranks, can a neighborhood tend towards an immune behavior that minimizes miscalculations? The prize could be the capacity for quasi-homogeneous fighting efforts against invading pathogens. Nature's policy in biology never has been to care for the individual's well-being; it has always been directed towards maintaining or even improving the genetic background of the species. To elucidate possible inter-human immune-links we must study living conditions at opposite extremes, for example, by contrasting comfort and well-being with overcrowded conditions in a prison cell. Although the latter is indeed shocking, *a priori*, it is to be expected that there could be some hidden benefit in this condition. What could it be? A certain fundamentalist view would point to the tendency of natural forces to maintain balance; for example, to play down the extremes from the Gaussian distribution curve, and it is likely that the benefit in this case could be fewer immune miscalculations.

Cognition is inseparable from alarm, a main means of defense. Unfortunately, alarm is also close to panic and self-harm. Therefore, cognition has to be severely filtered. We believe that people who are close are able to offer reciprocal support to each other, even in immunological matters.

Looking at human biodiversity, it seems appropriate to amalgamate the individual's immune structure with the surrounding milieu, and in this enlarged context, collecting and processing information widen the scope for investigation. Thanks to current acceptance of the apparently far-fetched butterfly effect, and the evidence for a widespread distribution of information *via* pheromones, it is becoming clear that conventional thresholds (of information-exchange) will have to be reassessed. That is why the possibility of a third immune medium alongside the conventional tissues and body fluids is being considered.

INTERDEPENDENCE THROUGH FAMILY BONDS

It is not necessary to go into detail about maternal ties, as they are shared by people in both the over- and under-populated areas of the world. The point is that antibodies that the mother delivers to her offspring, together with other immune-related molecules and some mononuclear cells, appear to amplify an apparent primary image of infection. In association with natural autoantibodies, the transferred immuno-active material may help the child to develop its immunological homunculus[32].

At birth, newborns receive—through materno–fetal blood transfusion—tiny amounts of the whole battery of mononuclear white blood cells of the mother. In addition, the mother may provide antigens and as-yet-unknown numbers of active molecules (as well as cytokines, hormones, neurotransmitters and drugs) across the placental barrier. To a certain extent, maternal immune support endures while nursing—the term 'support' denoting a wide-ranging immune interaction that affects the offspring.

Marital immune exchanges, principally through sexual intercourse, in a way also appear to be close to the envisaged community context—though there is no reason to believe that rural people refrain any more from promiscuous habits than do 'mega-city' dwellers.

Gene evolution influences group behavior. When siblings refrain from having intercourse, the reason is not purely because of taboo. Such behavior is shared by certain simians that live in small communities; when growing up together the males ignore their female siblings, even adopted ones. When they want sexual intercourse they leave the group and look elsewhere.

RESEMBLANCE WITH UNICELLULAR BEINGS

Microbes, even the very simple ones that lead an independent existence, when facing danger show a strong tendency to form colonies. When microbes form such compact masses it may become difficult to distinguish them from simple structure-lacking multicellular beings. It is noteworthy that many, if not most, of the molecules that are secreted by microbes appear to be able to communicate with other organisms—not only of their own species, but also with those of other species, even with far more complex organisms. The spectrum of information exchange ranges from rudimentary defense to symbiotic associations, some of them surprising in their sophistication. Microbes, when they switch from independence to living as part of a colony, reshape their defensive strategy. In a similar way, perhaps, when people become caught up in conditions of appalling overcrowding, hitherto overlooked factors may become involved to assist in bolstering up the defense. The small-scale and almost imperceptible piecemeal adaptation to the collective may account for it.

Survival over thousands of years under changing conditions should be reassessed. To a certain extent, humans resemble microorganisms in their ability to exist free or to assemble in colonies. The question remains, to what extent is human life in confinement able to change its immune machinery in positive or negative ways? Although the methodological difficulties of quantifying it are obvious, it is to be expected that, generally speaking, the ecological conditions inside a community are too feeble to help in providing protection. The incidence of disease through history, life inside prison cells, or, perhaps, life under extreme conditions may provide clues from which we can learn.

Community Immune Interaction

FIGURE 11 DOES THE MILIEU IN CONFINEMENT INFLUENCE IMMUNE BEHAVIOR?

If immunity only depended on some internal mechanisms of the host, why is it that healthy adjustments are found throughout the community? Ways to repel infectious challenges may differ, but often the influence of mutual hosts is involved. In order to avert danger, the individual's chances of getting help should be considered. Almost jointly with the advent of psychoneuroendoimmunology[33], researchers seriously examined people's interactions in situations of stress and confinement, thereby detecting immune repercussions[34]. Immune relations among subjects by indirect means are now being investigated in many centers.

INDIRECT IMMUNE TRANSMISSION—PLACEBO EFFECT?*

Interactive biological signaling has been demonstrated inside the individual, yet a causal link between psychological stress, immune function and disease awaits confirmation. The brain's limbic system† apparently assumes a central role in combining multi-systemic actions. The holistic psychoneuroendoimmunological approach may help to explore constitutive feedback loops for determining the dynamic behavior of the system. Stress is also involved. Potentially, it is able to affect interplay between the immunogenicity of pathogens and neurohumoral feedback. Positive or negative stress—depending on its condition being mild, intense, acute or chronic—can enhance or suppress immune responses[35]. It has long been a commonly held belief that social support—for example, the presence of a close companion—can lessen susceptibility to disease by dampening anxiety response[36]. Individuals with a strong social support network exhibit less morbidity/mortality[37], although this may be difficult to measure functionally. More objectively, a similar tendency could be shown in non-human primates[38]. In juvenile

* Placebo harks back to the 116th Psalm in the Hebrew Bible. Through a number of translating errors, the Vulgate Latin version came to contain the word 'placebo' (I shall please). Over the centuries the term placebo was applied to the Vespers for the Dead, derisively to servile flatterers and toadies, and to laments sung at funerals by professional mourners.

† Limbic system: term loosely applied to a group of brain structures (the hippocampal formation including septal areas, amygdala, etc.); involved with autonomic functions, certain aspects of emotion and behavior and with other activities.

rhesus macaques, Gust *et al.*[39] recently confirmed that the presence or absence of social partners modulates response to stress in the immune sphere. Lone subjects revealed a decline in helper T-cells and an increase in cortisol levels during environmental adaptation. Recent awareness of physiological changes related to eustress* is of special interest. By observing mice in the basal state and in response to exercise stress, Kingston and Hoffmann-Goetz[40] compared differences in environmental enrichment and 'housing' density, the former preventing negative stress-induced enhancement of T-cell mitogenesis by Concanavalin A†. Apparently, pleasant life activity induces eustress, thereby providing direct benefit to immune function; it also acts indirectly, by buffering immunosuppression that distress may cause. The buffer effect of eustress occurred irrespective of housing density. Eustress as a condition for good health is now widely accepted, and the immune system profits from it by mounting optimal responses[41] quicker, more vigorously and more efficiently. There is also a faster recovery to pre-stimulus levels (homeostasis).

It is clear that people interact strongly through psychological means when crowded together. Stress may act as a potent trigger and in such subjects it echoes through neurons, hormones and other messengers. There is now proof that the immune system also becomes involved. Ultimately, a mutual immune influence might result, obviating or supplementing other factors.

To appreciate the incompatibility between prejudice (almost always schematic) and complexity, we must remember that humans have survived—and often thrived—in the midst of unimaginable misery for over a million years. Until very recently, Man's main concern (and that of other species) was to overcome hunger. Hygiene is not a new idea—consider the Old Testament and the habits of the Ancient Greeks and Romans—but it was only put into practice in the last century after Semmelweiss urged hand-washing in order to prevent post-partum sepsis. Differences between confinement under extreme conditions, as was the case throughout history, and confinement, for example, during space travel, should

* Eustress: positive stress, physiological optimum, healthy. An enriched environment allows active entertainment to the animals, thereby providing eustress.

† Concanavalin A: T-cell mitogen.

not be underestimated. And their respective repercussions would, naturally, be different. The data from space travel have to be filtered out from space-related stress factors (weightlessness, radiation, etc.) and workload. Ancient chronicles mention people's astonishing resistance[42] to infection while clues from long-term (90-day) simulation studies on earth are still not clear[43].

In crowded habitats, inter-individual immune placebo effects should not entirely be excluded either. The placebo puzzle is persistent and widespread—for example, menstrual synchronization, or normalization of menstrual cycles triggered by intercourse. Often placebos provide straightforward help, but they have also been associated with all kinds of unpleasant side effects. Even 'placebo addicts', although very uncommon, have been observed. For an appreciation of the phenomenon, it might be apposite to imagine Sigmund Freud, in the 1920s, calming down someone's heterodox suggestion that the alarming mental problems affecting a patient could be corrected simply by regulating a disturbed thyroid function. Present psychoneuroendoimmunology has made us more flexible, although we are without an answer for all questions.

PARTICLES FROM HOST TO HOST

The purpose of the present work is to reassess the possibilities of interhuman immune cooperation. Data from the scientific literature do not really do justice to the extremes of the immune system.

Just as occurred at Auschwitz, and is still occurring in the shanty towns of today, attention is focused on adversity, ignoring anything that may hint at a positive side. We must ascertain to what extent overcrowded conditions hide the means to reinforce immunity. Disease transmission may not readily give up all its secrets, but perhaps subtle details will allow discernible features of cooperative immune response to emerge from the shadows of biology.

Our Past

History repeatedly refers to cities that went through a prolonged siege. Only a small percentage of the population would normally live in the congested fortified cities. Even under normal conditions the space available for permanent residents was insufficient—indeed, to this day many places remain where an average of six to eight people share a room. As agriculture used to be extremely precarious, everybody had to strive to fight starvation. It is hard to imagine the massive panicking flight triggered by the inexorable progression of Tito's legions. Although no exact numbers are available, there is no doubt that, during the Roman invasion of Israel, the crowd inside Jerusalem's walls increased substantially. Life

FIGURE 12 REGARDLESS OF THE INSIDE ENVIRONMENT, THE WALL IS THEIR LAST HOPE

inside the city would be almost impossible for us to imagine. With death all around, people were packed in close confinement and submerged in filth, piling up excrement and rotting remains, and deprived of water. Regardless of the consequences, starvation forced them to consume whatever was available, exposing them to havoc from pathogens or saprophyte attack. But, in spite of everything, defense did not completely disintegrate. The fighting was often prolonged. How did the individual's immune response cope? Certainly the particular environment required special adaptation. It is possible that sticking together immunologically turned out to be advantageous. One must infer that those who had no opportunity to co-adapt could not match the resistance capacity that the besieged acquired, infants and old among them. Any clear understanding of the diverse interwoven immune dynamics that evolved will remain elusive for the time being.

FIGURE 13 SLAVE SHIPMENTS PROVED TO BE WASTEFUL IN THE SHORT TERM AFTER ARRIVAL

Slaves' immune profile should do better than that of infants and old people, but collective immune streamlining requires adaptation.

Another fact from our recent past was the introduction of slaves into the workforce in urban communities. It was very hard for them to survive under the inhumane conditions imposed on them, with malnutrition, etc. as the norm. Chronicles tell that they often succumbed to infection, not infrequently with epidemic features. Mishap indeed came in a variety of guises but, for purely practical purpose, these people were selected at the prime of life—the optimal immune age. To mount an appropriate defense, the immunodeficient age groups (both old and young) of the stable population must have been at greater risk. Many historical accounts provide evidence that people were used to living packed together in filthy conditions, be it in towns, hamlets, or even in farms. It is conceivable that the masses who were used to these squalid conditions somehow became protected and the slaves' main handicap was their status as newcomers. They simply lacked the opportunity to learn to adapt to the particular environment.

COMMUNITY: AN INTRIGUING CONSTELLATION

The fact that the immune network has some means of being in touch with the outside milieu may contribute to a collective defense. The belief in the supposed boundaries provided by the skin may, in fact, be a relic of our earlier simplistic idea that an individual's immune system is unique to that individual. Since the immune network is in contact with outside milieu, it may contribute to collective defense.

Reference has already been made to *Schreckstoff* (German: 'fright substance') given off by agitated fish[44]. The chemical signal that alarms others of the species is detectable by them at extremely low concentrations; thus we are dealing here with solidarity in defense that is widespread in nature. To link such solidarity to a putative collective immune support system appears to be entirely appropriate. The response to sudden infectious challenges may vary between individuals, but often there is a mutual influence between them. It might be thought possible that, for the individual, the acquisition of information about invaders at an early stage may improve the capacity to counteract their virulence. To extract relevant information out of a 'milieu soup' presumably requires adaptation and time. In crowded conditions released molecules are supposed to remain active if the physical conditions (humidity,

FIGURE 14 ARE THERE MEANS TO READ OTHERS' IMMUNE RESPONSES?

temperature, etc.) are suitable. When those conditions are optimal, a bridging function might be imagined.

CHEMORECEPTION, PARTICULARLY VIA PHEROMONES

That the anatomical site of microbial invasion contributes to the intensity and duration of contact with the host's internal defenses has been accepted for a long time, whereas little attention has been paid to inter-host chemical connections in the microhabitat. Much has to be done to decipher physiological mechanisms of chemoreception and their role in patterns of accommodation to a hostile antigenic environment.

Chemoreception

A lot of the sensory processes to which organisms respond are induced by the environment, chemical stimuli being a main factor[45]. After an initial interaction with a cell receptor, before becoming converted to nerve impulses, the incoming stimuli are channeled through nearby cells, referred to as secondary receptors. By interacting with key cells, chemical stimuli provide much of the information about the environment. Materials, especially gases, may come a long way before reacting with so-called distant chemoreceptors, whose thresholds are usually very much lower than those of contact chemoreceptors. Among the behavioral results from distant chemoreception, biologists assign importance to orientation in a wide-ranging sense. Species-specific pheromones affect many aspects of life, such as territorial instincts, menstrual synchrony, etc. They have been shown to activate gene expression in cells from many kinds of animals, humans included, thereby inducing the release of the respective substances.

It has been held that invertebrate animals have only one chemical sense, with different sensitivities for various chemicals, whereas terrestrial invertebrates such as insects exhibit separable chemoreceptive capacities. Aquatic animals and terrestrial species with mucus-secreting skins are generally sensitive to chemicals all over the body. This sensitivity has been called the common chemical sense.

Common chemical receptors are usually free nerve endings (branching structures or dendrites). The common chemical sense is different from pain induction, and thus it is considered as a separate sensory capacity. Man and other terrestrial vertebrates have a remnant of this receptor system that responds to irritants in the mucous membranes of the mouth, eyes and genital organs. The subject is now under intense scrutiny, as diverse research efforts show. For example, chemoreceptive faculties among birds with developed taste buds and olfactory brain centers are being re-assessed. Previously these were assumed to be deficient and subordinate to the organs of visual and aural perception.

The complex story of biological means of interconnection has in pheromones a hallmark that encompasses huge numbers of species. These hormone-like, externally secreted molecules are remarkable for their diversity and in particular for their multiple functions. The list starts with olfactory perception, with which pheromones have long been associated. For example, honey bees scent-mark their own hive and areas around it with odors that uniquely identify that particular insect community for its members. In aggregating as groups or in dispersal, animals depend on their cognitive ability. Much diversity in pheromone-binding proteins (PBP) appears to have a filter function before receptor activation[46]. Usually protein kinases become activated upon pheromone stimulation. However, when several agents result in activation of the same transmitter enzyme, it is unclear how specific biological responses are generated. Among others, stress conditions or apoptotic agents become involved early on in pheromone responsive signal transduction cascades[47].

Transfer function probabilities improve in oceanic ecosystems, lowering threshold levels for encounter dynamics that prolong the distance over which the target (emitting) individual is perceived[48].

It is of interest to note that some pheromones act as important regulators of social behavior. They may be crucial among local populations in identifying members and excluding outsiders, with the concomitant possibility of links with immunity. Alarm odors alert ants, wasps or bees when colonies are endangered. Chemicals emanating from plants or from other environmental features guide many species, with, for example, toxic or repulsive secretions serving for defense. The conditioned stimulus-response cycle is common to many species. In the search for analogies with immune communication, it should be noted that pheromone-binding proteins appear to participate in chemosensory coding (cross-reactivity was not related to taxonomic relatedness of the species but rather to chemical relatedness of the pheromones to which receptors from these species were tuned)[49].

The idea that human pheromones exist dates back to the 1700s, although for long it was denied because no proven evidence for receptors could be found. Pheromones appear to be a link between nature and nurture in humans, with functions not restricted to the sexual sphere. Significantly, it was only recently that there was thought to be a reason to replace the term neuroendocrinology by psychoneuroendoimmunology. An involvement of human pheromones was suggested by an experiment in which women rated sweat samples from men. When the men's MHC was different, the women said that they smelled 'like sexy lovers'. But when the MHC was almost identical to their own, they smelled 'like fathers or brothers' and, in consequence, their rating as potential partners was poor[50]. Clearly, to find a distant MHC partner is in a woman's best interests. MHC can be seen as a genetic trait involved in species continuation with a teleonomic role that reaches to human immune behavior. Research into MHC molecules in the context of external immune links might also be of interest.

Regulatory Immune Cooperation through a Milieu Soup?

APPETIZERS: MHC classes I & II, peptides, IgA, IgG, cytokines, adhesion molecules, ancillary molecules, co-stimulators, soluble receptors, immune-related pheromones, enzymes, also mononuclear cells?

SPECIAL INGREDIENTS: Shared flora.

RELEVANCE: If environmental significance is discovered, it may become a factor in the putative immune-homeostasis.

Here we deal with a hitherto ignored immune habitat, woven thread by thread, surrounding the body. People who live apart from others are perhaps more at risk when they become exposed to hostile xenoantigens. It is possible that close urban contacts provide some sort of protective buffer effect, in much the same way that isolated microbes change defensive behavior when they become part of a colony. Overpopulation is associated with heavy pollution. The 'soup' that results depends directly on the population size, 'dirtiness' (odors, etc. as attractants) and degree of confinement. A shared habitat is fed by all the substances that the neighborhood indiscriminately sprinkles, sprays and oozes. If the group routinely feeds straight from a common pan or kettle, this also contributes to the effect. The convergence on a singular flora that derives from the different local floras and from attracting microbes from the surroundings, still awaits proper evaluation. The influence of pets must also be taken into account. There is no question that shared parasites have a role to play in the environmental immune constellations.

Conventional wisdom tells us that in biology every function requires a 'window' with threshold at the bottom and excess at the top. It has already been mentioned that there can be an immune modulating benefit from stress, providing it remains in the eustress zone. To become manifest, a collective immune behavior requires an appropriate density of molecules which can interact, but there are obvious methodological difficulties in quantifying this. The following all have to be accounted for: signal exchanges, mode of transport, antigens' receptors (concentration and configuration), and parameters such as inverse distance proportionality and respective receptor-capacity. Substances in transit often find their way to suitable targets, even deep-seated ones that might be supposed to be hard to reach.

It appears prudent to state that any drive to react immunologically takes more than a mere epitopic stimulus. It has recently been shown that the induction of the adaptive immune response might be enhanced by the actual inducer of acquired immunity, that is the antigen-presenting cell itself[51]. In fact, a human homologue of the Drosophila Toll protein, a receptor involved in the innate immunity of this invertebrate, has the potential to activate the T-lymphocyte response to antigen[52]. This makes it possible to invoke the concept of competition in the provision of information to helper T-lymphocytes or else between stimuli from both inside and outside the body.

Step by step, insidiously, the immune game rules may change by milieu-sharing. Implicitly, any mediator molecule carries a message. By gaining access to an individual, it might provide information about the emitter's immune status, for example cognitive assessment of current antigenic challenges. It seems that, in general, there is a lack of significance to the host. Apparently, connections between individuals via reciprocated signals become more significant the more polluted the shared milieu turns out to be. Through entanglement with the host's immune networks, a step forward in the adjustment of immune responses occurs, blurring the conventional boundaries that restrict the system to the tissues and body fluids[53]. Initially, it was thought that any immunological influence that adjacent hosts may emit could be disregarded, as it was expected to be well below the threshold level.

To assemble a functioning molecular constellation at the target with a critical concentration of each component seemed to be an improbable event. The fact is that some of the recipients' molecules act in support to allow even tiny external signals to trigger synchronized immune interactions. Most equivalent molecules from other species in the habitat can play a similar role. Furthermore, as modern physics, and chaos theory in particular, now questions classical threshold ideas, minuscule stimuli can no longer be automatically ignored. Such is the case when unknown numbers of molecules establish distant communications, for example, pheromones for sex, trail, alarm, or defense[54].

FROM HOST TO HOST

The incoming molecular flow is supposed to imply a potential to alter the protein composition of a cell's envelope, so that cells can activate or inactivate responses that otherwise would not have occurred. Proteins may act as apoproteins, that is as mere transporters for functional prosthetic groups, or perform mainly collateral activity such as, for example specific perireceptor functions. Chemicals are not necessarily valuable in themselves, suggesting an indirect effect by promoter molecules or promoter molecular fragments. An influx must give rise to just the required quantity of molecules at the right time. Often molecules are able to locate multiple targets and, in accordance with the amount and complementarities, interact with many kinds of functional molecules or molecular complexes, including those essential for immune reactivity control, with most prospective receptor partners inserted on the intensely active cell membranes. There is an array of molecular actions at receptor cells that still await convincing explanations. Among them may be cited: (a) chemical reactions at the cell surface; (b) adsorption of molecules on the cell surface; (c) penetration of substances into the cell; (d) enzymatic reactions at the cell surface; and (e) protein bonding in the cell membrane. Molecules acting to trigger events rather suggest a continuum pattern. Eliciting gene activation appears to be important[55]. Pathways relay direct messages regarding cell migration, differentiation, proliferation and, importantly, secretion. With some more detail, foreign signals' interactions with receptors (and multidimensional stimulation may be necessary) may modify cell activity, activating a variety of second messengers, including cAMP and cGMP, that involve the initiation by modulation of biochemical events required for cell proliferation, differentiation and function, thus indirectly influencing the production of cytokines and monokines[56].

For example, neurohormones and other kinds of molecules reach their targets *via* the plasma. Immune cells are provided with receptors for many neuroendocrine factors (ACTH, VIP, substance P, prolactin, growth hormone, steroid hormones, catecholamines, acetylcholine, and a number of HRF)[57] and Blalock *et al.* refer to opioid receptors in lymphocytes, granulocytes and monocytes[20]. Through the interaction of the immune network with messengers sent by other individuals, it is probable that there will be repercussions such as the following.

Cells have an ability to release all kind of substances. Soluble class I and class II molecules, often carrying suitable peptides, seem to be the most significant ones to be found in a 'milieu soup'. As antigens they can promote a range of host responses, from excess to tolerance induction. Peptides can also act independently.

Minitransference of immunoglobulin molecules may be important. Ancillary signals only seem to have an external role[58]. A number of them may have the potential to contribute to reinforcing or attenuating aggression. These signals also seem to be involved in the process of favoring or preventing tolerance. Much more information is needed for a proper appreciation of the extent to which adjuvant function endows meaning to invaders and to answer the question of whether milieu factors—as well as host derived ones—are involved.

It has already been said that cells can disperse integrating molecules. Notable among them are receptors that were thought to be firmly integrated into cells, membranes especially. Besides the possibility of being massively shed into the bloodstream, receptors can be identified in excrement, respiratory spray or in what simply oozes out. Receptors are only one example of the many bridging messengers.

DECOY - CAMOUFLAGE

We still lack information regarding aspects of decoy function and, in particular, the effects of pollution. Recently, Nydegger stated that decoy function is increasing and referred to soluble Fc receptors shed by platelets as a first line of immune defense[59]. Modern technology profits from similar principles; for example, fighter planes, by scattering aluminum particles, confuse radar systems to deflect anti-aircraft fire. In the same way platelet-receptors are shed into the blood-stream to deviate antibodies, thereby preventing damage to platelets.

FIGURE 15 A ROLE FOR THE COMPONENTS FROM THE HOST'S FLORA MIGHT BE TO DIVERT THE PATHOGENS FROM THEIR ORIGINAL TARGET

Research is under way that will soon provide a common cold treatment through the use of huge amounts of ICAM-1 molecules, which aim to interfere with the access of the hundreds of rhinovirus types to cells covering the upper respiratory tract. In the light of these developments we should look at possible decoy functions in the molecular and microbial repertoire of the concentrated milieu soup inside the human microhabitat. The concept of camouflage is deeply rooted in biology. Could it be that antigens combine with molecules significant to their functions? Many animals, plants and micro-organisms use camouflage as a weapon or as a defense. And from another perspective, camouflage is often a role for the host.

Antigens can be identified either nude or clothed. This means that in addition to nude heteroantigens in the milieu-repertoire, our immune system has access to shapes of jointly connected or perhaps integrated molecules that may act as supplements for response induction. Persistent recirculating antigens and minute interindividual antibody–antibody interactions could count in fluent exchanges of immune-cognitive stimuli.

At times, certain largely innocuous germs may react in an immunomodulatory fashion, possibly tuning in to a connection with flora and/or environmental adjuvants. Among the incoming signals we also have to consider microbes or retroviruses converted into vectors by attaching all kind of molecules including those of human origin. There is the likelihood that after customary exposure the host's behavior diversifies less.

A query in this context is to what extent anchoring external stimuli can trigger or otherwise modify functionally class I and class II molecules from the host—for example, obstructing the acquisition of tolerance, the opposite to previously cited suspicion of tolerance promotion by incoming MHC molecules. Usually their chemical architecture is specific in the extreme, minimal changes destroying or attenuating the molecules' properties. Rigid structural requirements, however, do not prevent identical incoming signals being the product of totally unrelated compounds. Pollution may offer such a possibility.

BOMBARDMENT

It is certainly important to see whether the reciprocal 'bombardment' to which hosts become exposed results in inconsequential or, on the other hand, significant consequences. To learn about this we should examine modes of living through history and of today's underprivileged people. Even if molecules only drizzle there should be some immune reactions (a metaphor, perhaps, for the Chinese 'water torture'). It is probable that a streamlining tendency will result, smoothing individuals' immune polymorphism. By contrast, it takes time for people not attuned to a particular environment to become familiar with a new milieu soup and its constant exchange of novel antigens and messenger molecules.

When considering the influence that 'dense milieus' may exert on individuals, inflammation should not be excluded. Wound healing is an inflammatory process that shares tools with immunity. As such, it seems to be an interesting subject for evaluation under changing inter-host connectivity.

Tissues, Body Fluids and Polluted Air

Bridging between immune systems: a malleable collective immune force?

While the microscopic pieces are becoming clear, macroscopic behavior still remains a mystery. Attention has been directed to possible ways for incoming molecular signals to establish links with the internal immune network and to become involved with adjustment of response quality. The fact is, that by gaining access, mediator molecules could provide information about the immune status from fellow people, perhaps cooperating with cognitive evaluation of antigenic challenges. Bridging between immune systems would be in line with the standards of behavior in nature. To be sure, it has to be acknowledged, that players from outside are by no means the only ones under suspicion of influencing immune behavior.

The immune transmissions that we attempt to approach require a fluid-conducting material. If, in effect, functional heterologous molecules can have a role, we should consider the idea that, potentially, immunity may expand to encompass more than body fluids and tissues. A polluted 'milieu soup' with sufficient density could become a minor immune medium, intervening to adjust host response rates.

Community–immune features apparently develop (result) from substitutions (accumulated?) of immune response priorities (selection). As I have already pointed out, like conventional senses, the 'immune sense' routinely adapts to a restricted selection of stimuli. So the questions arise, as before, what factors are important; are they shared; at what levels are pathogens repelled; and what is their route of transmission? Here, comfortable living conditions are not examined; the focus entirely centers on people with no option other than to stick together, confined in unhygienic conditions. Their habitat will get polluted in many ways by interacting molecules from hosts and a particular microflora that results from a convergence of the flora from the surface of the body and its orifices. It is already traditional wisdom to consider the normal flora constellation beneficial, and it is possible that it provides a kind of bottleneck for invaders to overcome, thereby enhancing the immune profile's rejection capabilities. To the invading organisms, competition for space may pose a serious obstacle. The apparent equilibrium in a habitat does not prevent continuous in-fighting. In spite of putative benefits, alertness remains justified because of ever-present ambiguity. So there are many factors to be considered, often acting in combination. And, over time, interactions will unavoidably produce immunological repercussions. Large cities and towns not only concentrate people, but also animals—in great numbers and densely packed. Human habitats attract micro-organisms, insects and other animals. Most of them are harmless but the question remains whether the habitat's 'molecule traffic' may result in gain or loss to the defense of the hosts.

Anamnestic phenomena* constitute an exciting research target in immunology. They are possibly involved in the changes that longstanding confinements induce. A tendency to erase individual immune differences could be another aspect of habitat sharing. It is an appealing idea to consider that crowded living conditions could strengthen defense in infants between the time when maternal immune-support fades and their own immune machinery fully develops. A putative milieu memory could also mean less disease-susceptibility for old people, and perhaps implies attenuation of aberrant immune behavior. Any community protection for the weakest sections of the population would sharply contrast with 'disease-proneness' of people who lack adaptation to the habitat.

A key question emerges—could a contaminated human milieu entangled with individual networks, as a whole, supersede individual solutions when people 'learn' to integrate?[60] Biological evolution promotes networks throughout the body. A strict adjustment to conventional anatomic limits does not appear convincing. We have to learn how the immune network spreads beyond the bodily limits and how it assimilates external influences to its functioning. What role do traditional microflora play in evolving as part of the milieu?

* Anamnestic phenomena: (Greek *anamnestikos* = easily recalled) used for immunologic memory that boosts the response by re-exposure, often surpassing the previous ones

Stimuli, although variable, may maintain certain characteristics over time, and reinterpreting them could point toward the blueprint of common defense behavior. Ideally, immune response should combine collective and individual patterns and the point is to establish the dominating factors in the putative interplay. Responses may not be simply 'specific', as circumstances change. To complex antigenic provocation each individual, rather, is an 'open-ended' responder. Immune defense might be strengthened through immune molecule traffic but at the risk of interfering with its function.

An ecological approach is at least a two-way street, a subject that has to be seen as a complex whole. In any habitat all information sent out or received supposes the existence of emitters, and subjects that receive messages from their neighbors are supposed to send out their own.

INDIVIDUALLY VERSUS CONCERTED

Free-living individuals react far from stereotypically, and with wide-ranging responses. To lose the status as an independent immunological entity is, perhaps, fundamental to life in overcrowded conditions and a kind of homogenization is associated with any conversion to communal living. It happens to children at school as well as to citizens in the army—but in overpopulated prison cells things are much worse. As distinct from eustress, prison inmates are under intense negative stress, and it must be assumed that inmates' synchronization also involves the immune profile. Substances that fellows exchange combined with factors from the habitat may act as regulators. In the confinement context, not only does the persistence of contacts and exposure to the habitual stimuli have to be considered, but the impact of new ones must also be taken into account. The challenge is to establish the scale on which, for example, droplets of diverse nature transmit factors that improve or worsen the course of diseases. If it turns out that habitat proves to have a potential for tuning the immune system, it might have been significant for human survival over millennia of instability. So far, standards for collective immunity have not been demonstrated, so here we must look at the extremes. Any understanding of life-style-dependent differences in immune responses would shed light on the elusive role of the environment in the spread of disease, and hosts' resistance and adaptive behavior.

People, like free-living micro-organisms, seem to depend on themselves, unless they get signals from others with whom they are in contact. To what extent they depend on such signals remains to be seen. If the proposed theory turns out to be correct, one could say that, in terms of habitat, sticking close together implies a kind of ecological immune communication among people. In order to influence the threshold of disease transmission, perhaps we should consider a synthesis of bad and good results, a 'spectrum' of consequences.

We conclude that reciprocal signal stimulation should affect immune behavior. For example, it is possible that the most immune-experienced members of a vitiated habitat, by releasing suitable substances, could carry out successive steps on the route to converging immune response developments, even to initiate synchronous responses. To build common experience into the image of infection seems teleonomic*, as it has to be seen as contributing to basic success in a biological sense. The framework it provides supersedes individual solutions pointing to 'emergent supra-clonal attributes'. Immune interplay may perhaps become rewarded by practical repercussions in the context of weight and resulting bias. Partnership under extreme conditions may be life-affirming and could prevent the spread of disease. Immune resistance should not be underestimated as a simple consequence of exposure, as its comprehensive role in species' survival reveals.

* Teleonomic, Gk. *telos*, end, and *nomos*, law. The doctrine that life is characterized by endowment with a purpose, or that existence in an organism of a function implies that it has had evolutionary survival value

On the Immunological Homunculus and an Attempt to Redefine the Immune Sphere

Irun Cohen had good reasons to assume that the immunological homunculus mainly expresses some of the host's long-standing antigens[61]. By denying that anatomical body limits are absolute, the possibility arises that close substances may contribute to the virtual 'immune ego'. Mainly we think about flora components, those normally present or those resulting from convergence. If true, the

FIGURE 16 EBB AND FLOW: DOES THE MOON'S EGO TOUCH OUR SEAS?

immunological homunculus would not be symbolized any more exclusively by selected host-specific molecules. The 'homunculus' notion was introduced by neurologists to outline certain neurons normally related to 'self'. The facts that the notion of self is relative and that it also depends on external factors were exploited by the Nazis, as can be seen from the damage they inflicted on their victims' egos by ordering them to shed their clothes.

Lessons from defensive cooperation among micro-organisms (for example, the spread of antibiotic resistance) have also led us to question the current immune-centric view. Bacteria improve their defenses when they switch from isolation to integrated colonies. As we have shown, in isolation, humans behave totally differently to when they are forced to integrate by confinement, in which situation, evidently, exchange of messenger molecules manipulates the environment. 'Immune talk' ranges from a mere hint to a system of great complexity.

SKIN

Many immune responses integrate throughout the organism with involvement of the external body surfaces—gastrointestinal, respiratory and genitourinary tracts among them—normally maintaining their respective associated microflora in harmony.

Skin, indeed, encompasses a coherent whole with autocontrol and important defense functions. There is no question that skin and the mucosae of respiratory and gastrointestinal tracts provide the first line against invasion. The keratin-producing outer layer of cells efficiently covers skin surface, and glands cooperate by secreting oleic acid, lysozyme, etc. that can destroy bacteria. In the more developed vertebrates a great variety of immune elements cooperate. Although less tough than skin, gastrointestinal, respiratory and genitourinary mucosae also act as an effective mechanical barrier to infection. All the covering cell linings are constantly being renewed from below. Endothelial cells have cilia to expel mucus that specific cells produce and which traps potentially harmful particles; it also facilitates the passage of food. In addition, mucus concentrates IgA immunoglobulin.

Many released cell components, at the various orifices of the body, interact with the external contact sphere. The body surface has many options for interactive communication, favoring the notion that the immune network extends beyond conventional limits.

LIMITS FOR ASSOCIATION

That the environment contains many potentially destructive microorganisms is obvious from the speed with which animals decay after death, although many microbes live harmlessly in contact with their hosts. Normal microflora are abundant on skin and in orifices and tracts—the important point is that the inside of the body remains sterile, and the immune system acts to this end against any invader, whether it is breathed in, penetrates the skin by punctures or cuts, or gains access by any other means.

The relationship that pathogens and victim establish is not exclusively one of predatory harm. It may affect the whole immune equilibrium of the host, with the involvement of other networks in the body. With regard to repercussions beyond the limits of the skin, exact information is lacking.

It seems likely that immune exchanges between the individual and his surroundings are a constant event, although for practical purposes the effective radius of action appears to be extremely restricted—the reason why it tends to remain mostly unnoticed. Nevertheless, over time the milieu might exert some influence on the immunological autonomy from nearby members of the community. Bearing in mind the importance of ecological links in that which concerns immunology, it may be a two-way street. Other questions are: can the external human environment contribute to the image of 'self'? Is it biologically acceptable that identification and defense exclusively center in 'self'? As yet there is no conclusive proof that immune-subjectivity exclusively belongs to the individual.

MICROCHIMERISM THROUGH CONFINEMENT?

A similarity to the diverging ways of human life under discussion suggests that microbes have the option to exist independently or to gather in colonies that are sometimes difficult to distinguish from multicellular organisms. The same question arises here—can the environment influence the colony by impeding or by promoting the acquisition of reciprocal immune tolerance?

In recent time much has been learned about microchimerism as a result of blood transfusion. It is surprising how often donor cells survive, sometimes being detectable in the host for a long time[62]. It should also appear to be feasible that cells get transferred by means of a milieu soup, and retain the ability to remain alive in the process. As has already been hypothesized, alongside body fluids and tissues, the habitat enclosing them may become a minor immune medium. A study design for hypothesis generation has been made. In the first place we think about babies and immunodeficient subjects inside shanty towns. At first sight, the simplest way to get an answer appears to be by looking at statistical differences in transfusion intolerance under sharply contrasting living conditions—that is to say, normal dwelling versus crowded.

FIGURE 17 GUESS WHICH ARE THE ELEMENTS FROM THE COLLEAGUES WHICH MAY CONTRIBUTE TO OUR IMMUNE NETWORK

Unfortunately, data are not yet available. There are too many factors that require a careful evaluation. Standards of transfusion intolerance, for practical purposes, remain far from consistent in different parts of the world. The gap between so-called developed and developing worlds' blood-banking is unacceptably large as regards material resources and trained personnel, and especially with respect to clinical assessment. It is still almost impossible to verify if transfusion symptoms simply derive from pyrogens or if the cause is a specific sensitization against blood cells (any kind) or against a plasma component. Furthermore, in population clusters, the heterogeneity of MHC molecules differs dramatically. Finally, conditions such as climate, nutrition and others have to be taken into detailed consideration.

Autoimmunity – a Private Condition?

Healthy individuals have a repertoire of autoreactive antibodies and T-cells which mainly originated early in ontogeny. Physiologically, they are supposed to be involved in immune homeostasis and self-tolerance regulation. Still, often the transition to pathological autoimmunity is misleading. We should remember that, generally speaking, target antigens are dominant. Autoimmune disease *per se* tends to be considered as an immune deficient condition but frequently appears to be a by-product of immune deficient disease. Stewart *et al.*[63], by referring to computer simulations of the immune network, show that the greater the degree of connectivity of a clone, the greater its tolerance to chronic antigenic stimulation. This tolerance does not correspond to an absence of response on the part of the system as a whole. On the contrary, stimulation by a 'tolerogenic' antigen results in widespread modification and overall activation of the whole network. This suggests that on an autopoietic view of the immune system, autoimmunity arises not because of the presence of self-reactive clones, which is completely normal, but because such clones are inadequately connected to the network. While not all individuals with immunodeficiency develop autoimmunity, nor all individuals with autoimmunity are deficient, defects within certain components of the immune system carry a high risk for the development of autoimmune disease[64]. Although there is ample evidence for a genetic role in autoimmune disease, only 34 percent of monozigotic siblings from patients with rheumatoid arthritis will become affected by the disease; and in no autoimmune illness does the percentage surpass 70 per cent. This might indicate that, in many cases at least, environment takes part in the development of the disease[65].

Autoimmunity can be seen as a consequence of an internal storm in the immune system. Although in many cases it has become clear that autoimmune disease does not depend exclusively on the patient's own state, to think about the outside environment as directly influencing the condition is almost certainly an oversimplification. To evaluate clinical evolution, rather than particular autoaggressive antibodies or cells, a complete answer may provide common immune features, provided they become properly identified. Often, a certain infectious agent is relevant[66]. In addition to heteroantigens, superantigens may also act as a trigger[67]. All in all, autoimmune disease is not necessarily a strictly personal condition; the environmental status should also be examined, especially when the public habitat is mixed. However, much has to be learnt about how autoimmunity turns into immuno-pathogenic self-aggression, and about the whole panorama of flare-ups and remissions contingent upon that state.

AUTOIMMUNE BEHAVIOR

Like many aspects of an individual, where careful observation allows identification of characteristics that may be unique and can define specific behavior, the immune condition is no exception. Repeatedly we are asked to recognize a special immune behavior. Autoimmunity manifestations in a patient often 'run to form', but this is not always the case; for example, in acquired phospholipidic syndrome the expression of the condition may be recurrent abortion in some patients, while others tend to venous thromboembolism but no arterial disease, and vice versa[68].

The flare-ups and -downs that characterize the symptomatic course of autoimmune disease point to changes in the individual's immune behavior. Stress may exert an influence[69]. It would be interesting to know if cramming people together is able to exert an effect on any of the manifestations. Autoimmunity is often associated with advanced AIDS and, to a lesser extent, with earlier stages of the disease. In patients from the Francisco Javier Muñiz Hospital in Buenos Aires, where terminally ill AIDS patients are treated, autoimmune manifestations are frequent[70]. The huge hospital for infectious diseases covers an area of 8 hectares, and contains a prison building for convicted criminals who have AIDS. This provided a possibility of comparing outpatients, hospitalized patients and prison inmates at similar stages of the disease. In response to my suggestion, González and his team hastily collected data on immune anti-erythrocyte antibodies, direct antiglobulin test (DAT) positive, in the three groups of AIDS patients, all of them belonging to stage IV-A to IV-D (classification proposed by the CDC[71]). Interestingly, González observed highly significant differences in DAT positivity frequency among the patient groups. Prison inmates showed the highest incidence and outpatients the least, while the hospitalized patients' group values were almost equidistant in the middle of both extremes (personal communication). Every day we witness in the media the damage inflicted on victims of assault who fight back. The immune response reacts similarly to shock, and the community can have a collective defense mechanism also. Unfortunately, the hospital, overwhelmed by around 150 000 admissions a year, is in bad shape and with chaotic treatments. Their failure may be attributed to the patients' lack of cooperation, irregularities in drug supply and follow-up, and poor attention to the many life-endangering complications. High mortality is a consequence. The as-yet-unpublished results show convincing evidence of a three-stage DAT positive incidence in accordance with the degree of crowding. Obviously, there are multiple factors to be considered and much discussion is under way; for example, there is still no conclusive agreement as to whether current antiviral treatments modify autoimmune disease incidence in HIV-infected patients. Our special interest concentrates on the apparent parallel in DAT positive incidence-AIDS and patient crowding in the advanced stages of AIDS. In other words, anti-erythrocyte antibody detection increases, the more patients crowd together and establish physical links (although here we omit consideration of sexual conduct). It would appear that one can make a distinction between the effects of permanent confinement (as in, for example, a prison cell) and those of optative confinement (as in, for example, shanty towns) on a person's immunity independently of their own immune condition.

POPULATION DENSITY

If epidemics are examined, interestingly, statistics do not prove that crowded living means increased disease incidence or mortality. Widely scattered farmers are not better off, contradicting the usual assumption that life is healthier in the countryside. Again, it is tempting to infer that when confined in insufficient space, permanent and multiple contacts reinforce immune protection. A noteworthy characteristic in biology is self-sacrifice as a means of defense.

Globalization

There is nothing new in the notion that all creatures reveal interdependence patterns, but now the condition is accelerating with no end in sight. Before discussing present anxieties, some anecdotal aspects may illustrate my points.

Wind: organisms, mainly microscopic, but including spiders, mites and insects have been collected on flights up to 3000 kilometers from land.

Water-flow: many insects depend for food upon the organic matter drifting from the upper levels of a river.

Ocean: not only marine invertebrates, but also freshwater animals, passively cross oceans by drifting with currents. The odds are considerably against terrestrial organisms surviving the ordeal, but there are exceptions. In 1892 a remarkable floating island, 30 meters square, was observed in the Atlantic with trees several meters high, having drifted not less than 1600 kilometers.

Attachment: animals converted into vectors can disperse organisms over huge areas. For example, mosquitoes have been found to carry a virus many kilometers before infecting a host.

FIGURE 18 AN UNFORGETTABLE INHERITANCE OF THE HISTORIC ENCOUNTER

Excrement: this widely disperses the seeds of some plants. Many other mutualistic associations contribute to dispersal.

Travel: long-distance flights and shipping of people and objects are the main culprits of globalization.

It has to be acknowledged that humanity has been dealing with globalization for a considerable time. The difference now is that contacts are more-or-less simultaneous across the entire world. The expanded range of contact implies much more severe demands on immunity. Since everything is in flux, variables and fluctuations are unclear. Nevertheless, as long as scientific approaches can be perfected, the overall long-term outlook does not seem too bad.

The scope is becoming continuously broader as, increasingly, biologists and ecologists become convinced that our globe functions as an integrated network. The immune function, that practically encompasses the body and its physiology, may share in the planet's evolution, from incipient connections with the immediate habitat to the possibility of expanding to more remote networks. We can, perhaps, draw parallels with the popular view of Gaia as an organism.

FIGURE 19 GLOBALIZATION: PRESENT

FIGURE 20 PLAGUE

FIGURE 21 WHAT SCARES YOU? THE COMPUTER VIRUS OR WORLD-WIDE EPIDEMICS AROUND THE CORNER

AWAITING AN OPPORTUNITY

Most of the interdisciplinary tools that have been outlined in 'Background' apply to epidemics. Clearly, the subject still causes extreme puzzlement. Sensitive dependence on initial conditions seems an obvious factor but the early traces in the development of an epidemic disappear. As a result, we are helpless to answer the easiest fundamental questions about the future of a plague. The imprecision built into the calculations rapidly takes over because of the relationship between the particular responses by different hosts. We have to consider environmental disturbances of diverse nature and, of course, the many pollutants, largely stemming from consumerism. Many are the factors that alter the normal physiology of pathogens and host–pathogen interactions. Observations have indicated that a bacterium and its host stimulate each other to carry out successive steps in a way that regulates the development of both organisms. People often have to deal with bacteria that apply particularly clever tactics and the question remains how to fight them off. Variations in habitat and living conditions, naturally, matter. To follow the course of an epidemic, population density and hygiene have to be considered first. Often differences in the disease's course seem to depend on contradictory signals. The chaos theory brings immune behavior with its unforeseen changes into the limelight. Dependence on initial conditions, minimum thresholds, etc., are increasingly being investigated, encompassing a spectrum that runs from firebrand-like progress to complete reversal of the scourge. As has already been said, overcrowded living conditions pose questions that should be reassessed. Disease spread, contagion, and recovery pattern are forever changing, thus hampering any informed prognosis. As a consequence, so far, no general agreement for the optimal assessment of the progress of epidemics has been achieved. Here we outline views from two currently diverging authors, Ewald[72] and Wills[73].

Wills sees plagues as temporary blips in an otherwise balanced natural landscape.

> 'Plagues that afflict us do so because we have upset the balance of nature. The conditions that engender plagues always end sooner or later. This always means that plague pathogens will not survive in the long term unless some of them manage subsequently to lose their virulence and revert to the habits of their low-profile relatives.'

Ewald vehemently opposes these ideas, adducing that Wills, by defining plague as something that eventually subsides, ignores diseases that continue to affect large numbers of people for millennia. Ewald further states that plagues that involve the presence of pathogens in a large number of hosts may taper off in a way that is unrelated to any parasite selection for lowered virulence.

> 'Often plague burns out because enough hosts die so that their population decreases and the pathogen can no longer be transmitted effectively. If the selective pressures of the competing strains favor the more vicious variants, however, the pathogens may maintain high virulence. A low profile may result from a low rate of infection rather than the loss of virulence.'

Some attempts to generalize in epidemiology seem to lead nowhere. So we should not expect for each plague a similar long-term evolution. Such is the case of parasites like the *Plasmodium falciparum* that vectors, anopheles mosquitoes, directly introduce into the bloodstream, thereby providing an entirely passive access to the host. As no hurdles have to be overcome for penetration, malaria offers pathogen varieties fewer chances to prosper. They simply have no need for versatility. A tendency for change results from the fact that, as plasmodia dislike mutated hemoglobin, the high incidence of thalassemia in the Mediterranean Basin is seen to be caused by the decimation of people with normal red blood cells. Very different is the lot for many microbes to which such a direct access is denied. They have to fight and force their way into their prospective host, a process that often is only achieved after fierce competition with resident flora strains. A picture emerges that reminds us of bullfights again. There are even more schematic points of views, as for example the one worked out by Cherkasskii[74], which he presented in ecological terms. By analyzing an epidemiological idea that is based on the principles of the theory of systems and the theory of information he concludes that the epidemic process is organized by the same principle as living matter with a kind of stability that self-regulation ensures. The author also estimates that human life conditions have been shown to be organically incorporated into the structure of the epidemic process as a regulating subsystem on the socio-ecological level.

The prevalence of epidemics is intimately linked with the adaptive capabilities of species. Here the Darwinian approach must emerge. Transformations that may translate to the epidemics genius are much favored by ephemeral life spans of the pathogens[75]. On a lesser scale, the reader is already familiar with the idea that immune defenses actually constitute an important part of perplexing intricate networks. Altogether they affect the pros and cons of contagion. The menace inherent in epidemics[76] subjects the immune system to pressures beyond the norm, thereby forcing the individual to seek adaptation or, to put it another way, to learn. It must be suspected that in the midst of epidemics, people may develop something like immune-synchronization—an unknown relationship that, teleonomically, tends to revert the (algebraic) sum of the facts from the resulting 'reds'.

According to our view, when habitat becomes overpopulated, immune network entangling is a possibility. Signalling between individuals may be widespread. The milieu soup, to become workable, must be critically polluted by a mixture in which trafficking molecules from hosts and shared microflora, possibly in some sort of balance, predominate. No doubt much depends on the nature of the usually pathogenic intruders. Should any beneficial effect show up, infants' immune behavior might be more receptive to environmental constellations. It sounds plausible that, by signalling appropriately, immune bridging between fading maternal protection and full-blown adult immunity could result. According to our (unfortunately non-professional) survey, during the usual winter epidemics, children from shanty towns do better than the more privileged ones at resisting infection—however, when the shanty town youngsters do get ill, their handicap as underdogs shows itself by a much worse disease course.

An understanding at the molecular level of the intervening period (interregnum) provides an interesting perspective.

Ambiguity in Aggression, Tolerance and Immunity

FIGURE 22 THE OSCILLATION PATTERN

Yin-Yang expresses the oscillation pattern. Yin-Yang are the two opposed and complementary principles that make up everything. Yin and Yang are the dualistic manifestation of the one and only eternal, indivisible and transcendent principle: Tao.

Leaving aside the nostalgic symbolism expressed by the caption, the scientific establishment is well aware of the subject's significance. In nature, no manifestation might be entirely exempt from ambiguity. Ambiguity never concerns a single entity. It is the relationship between things that, forcibly, implies ambiguity. That relationship is not a static notion, it suggests options—pros and cons. Bearing in mind interdependence up to a certain degree, any behavior will be shaped by external influences. Think about a river that divides into different streams and then becomes one again; similarly, light waves have a potential for such behavior. In biology, ambiguity is omnipresent, occasionally only as the tip of the iceberg. Immune-status becomes involved as changes and not infrequently has connotations of ambiguity. It should always be kept in mind when wavering between enhancement and suppression, such as the often surprising progressions or regressions of autoimmune disease.

By developing the interactive immunity subject, a distinct normality tends to evolve amidst promiscuity, perhaps even coming close to equilibrating a chaos-like state. And it is ambiguity that emerges. On the other hand we are dealing with uncertainties in the conduct of supposedly sympathetic microbes that belong to the host's flora. This is the case when, for example, the host's immune system weakens because of disease, old age or stress. The escorting microbes, that for long proved to be beneficial, exploit the opportunity by turning into potent aggressors, not only against the host, but—as their potential to spread may complicate things—potentially the entire community. Problems may affect the whole microflora network by expanding beyond its habitat to orifices, surfaces, etc. If the outcome does not become fatal, ambiguity acts here as well, but because of ambiguity, huge problems may arise.

It might be apposite to say a few words about viruses that have remained harmless or latent for millennia. Even after such a long period, and perhaps because of environmental changes, there is a possibility that the viruses switch from innocuousness to becoming the trigger of an epidemic. Such a potential sword of Damocles can mean disaster for billions of people; to keep an eye open for ever-present/latent ambiguity is indeed of paramount importance, and modern man should not consider himself safe from the dangers of plague—nor should he suppose that that danger necessarily comes from afar.

FIGURE 23 CONNIVANCE

FOCUSING HOST-ASSOCIATED MICROFLORA AMBIGUITY

Immune responsiveness may be become activated or inhibited, and some factors may have either effect, depending—among other things—on their specific immune function and the state of the target cells. The subject is complex, going way beyond the scope of the present work.

Existing theories on the role of the environment do not take into account the diversity of opportunistic bacteria that are found in a variety of strains within the same species. The recent emergence of *Escherichia coli* 0157:H7 is a reminder that bacteria may reinvent themselves, adapting to new hosts, new conditions, and new antibiotic countermeasures; these types of bacteria are known as mutators[77]. *E. coli*, some strains of which are essential inhabitants of the intestinal tract, may occur under the form of the highly pathogenic K1 serotype, K1 being nothing but a conglomerate of sialic acid making this germ capable of escaping complement activation[78]. Responsiveness can be modified over time. For example, in HIV

FIGURE 24 UNITED EFFORTS ARE WELCOME, BUT ASSOCIATION IMPLIES RISKS

infection, independently of the consequences induced by antigenic mutations of the retrovirus, changes in hosts' immune responsiveness play an essential role in determining disease progression. Accordingly, the outcome depends on a whole spectrum of antagonistic and synergistic variants.

The potential for ambiguity in species that belong to the normally associated microflora perhaps best reveals itself in the *Staphylococcus saprophyticus,* a coagulase negative organism of the normal skin flora. As the name clearly indicates it was previously considered nonpathogenic. The condition changes when the microbe becomes free from a deficiency in the host's immune system. It is now evident that *Staphylococcus saprophyticus* is the main cause of urinary tract infections. Indeed, it is second only to *Escherichia coli* as a cause of cystitis in sexually active young women. It is also commonly found in elderly men who have undergone urinary tract operations. Systemic bacteriemia is a rare event when upper urinary tract infection occurs.

Since the origins of humans, their life-cycle has been exposed to multiple diseases. Humanity was able to overcome an immense

FIGURE 25 IS IT WISE TO DISMANTLE SUPPORTING FLORA?

variety of micro-organisms in their diverse stages of development and the ways they mutate and evolve—but, in the reciprocal relationship, hosts are often taken by surprise, revealing invaders to be ubiquitous and complex challengers. It has already been mentioned that ambiguity matters in species' ecology and that not only obligate pathogens, but also the whole spectrum of innocuous-seeming saprophytes have the option to become pathogenic.

IS THERE A ROLE FOR 'PROMISCUITY'-ASSOCIATED MICROFLORA?

The currently idealized concept of bodily hygiene was exported by US troops during the Second World War. For many years, the Europeans only reluctantly incorporated it into their living habits. To attribute a supportive role to the usual microbes that populate skin surfaces is, at least, questionable. It is possible that the Americans show more vulnerability to skin infections, but, for practical purposes, the annihilation of skin flora by excess washing is clinically irrelevant, although stripping away important immune-related molecules and locally secreted protective chemicals can translate into unwanted consequences. We have seen that even the *Staphylococcus saprophyticus*, in spite of its name, converts into a risk factor in the elderly or when the host's immune defense is otherwise debilitated.

A word about the paradox of excessive intermingling: whereas rural dispersal exposes people to less disease, the absence of challenges increases their vulnerability. In effect, isolated individuals seem to be more at risk from hostile heteroantigens. Seen from this angle, although there are, undeniably, negative aspects, indiscriminate intermingling also seems to have the advantages of a more extreme immune interactivity among people (and perhaps among pets as well). In the future, certain peculiarities may be related to the environment's chemical changes by virtue of such mixing. The question as to whether indiscriminate intermingling exerts an influence on the mutation rate of antigens as yet has no definite answer.

Antibiotics and Pesticides

FIGURE 26 ANY RESEMBLANCE TO ANTIBIOTICS' FATE IS A MERE COINCIDENCE

The increase in antibiotic-resistant infections and the shortage of new breakthroughs in the field have inspired our cartoon showing the expectations of retired people versus the ability of the shrinking workforce to afford to maintain those expectations. The plethora of

FIGURE 27 FOR OVER A MILLION YEARS HUMANOIDS SHARED THE GLOBE WITH MILLIONS OF SPECIES. DOES OVERWHELMING POWER IMPROVE THE OUTLOOK?

antibiotics, at least for now, is failing. For example, salmonellas exhibit a high rate of plasmid-related antibiotic resistance. Often a transposon encodes a b-lactamase that is inserted into either a plasmid or a bacterial chromosome. These self-replicating circles called plasmids are best known medically for their ability to confer antibiotic resistance on host bacterial strains and to transfer this resistance from one bacterium to another. Multidrug resistance may result and has already posed serious problems even in hospitals in the developed world[79].

Our cartoon is losing credibility, however. We are in the middle of a profound revolution concerning pesticides and anti-infectious agents, and the whole panorama of pests and fighting infections may well soon change. Through our expanding insight into molecular biology we should be able to switch from established paradigms and to convert predatory methods into friendly biotic solutions to avert disease from plants, animals and man.

The reason that Quattordio's picture has been retained is that, even in today's climate, long-standing methods with huge support from established interest groups cannot easily be dismantled; considers, for example, the tragedy of tobacco-addiction. There are many doctors who persist in indiscriminate prescription of antibiotics and, worse, they teach the habit to their students. In addition, a significant proportion of veterinarians and farmers are doing no better. So, it is our feeling that the cartoon and a few sentences should be left in place to help counterbalance putative negative influences on the student.

We learned from acquired drug resistance that collective defense works even at the unicellular level. It has been shown that certain microbes, when they manage to survive a 'fool-proof' new antibiotic, massively shed mutated molecules which are incorporated by others of the same or different species. This has the potential to annihilate therapeutic efforts in hospitals.

In spite of the many triumphs of the antibiotic era, it cannot be denied that it is not possible to eradicate disease by the action of antibiotics. Doctors who approach public health interests in a responsible fashion contrast sharply with their colleagues who simply go for the 'quick fix', heedless of the immediate or long-term consequences. A total absence of planning is not uncommon. Doctors also may have political interests, or interests in drug trials. The obvious conclusion is that far-sighted behavior simply is not normal. As the Bible puts it, prophets were often lonely.

There is no overall agreement that the antibiotic era has been counterproductive. Even when treatment proves to be only partially

FIGURE 28 ANTIBIOTICS AND PESTICIDES EUPHORIA IN RETROSPECT: MORE HARM THAN GOOD?

lethal to the microbes, it may enhance the host's immune system. Some protocols are at present running on this premise; for example, the treatment of HIV-infected patients. Nobody believes that antibiotics are immune-stimulators; on the other hand, any debilitating action upon micro-organisms gives way to a proliferation of the survivors, in particular those that had mutated to resist the antibiotic.

The following example illustrates what may happen in future: even if an insecticide is harmless to people, it may destroy predator insects that—in nature—would keep pests in check. Any predators that gained resistance to the pesticide may survive its action, and such superpredators could be reared and distributed together with the pesticide.

STARS' ROLE-UNCERTAINTY?

When, under natural conditions, species grow out of control, the contemporaneous equilibrium tends to become destabilized. With regard to humanity, this principle applies not only to the plagues that repeatedly decimated it, but it also applies to the rise of might.

FIGURE 29 LONELINESS

When in the 1950s antibiotics became popular, optimists uncorked celebratory champagne as 'masters of nature'. But what is thought of as fundamental can change. As has been mentioned, increasing failure in conventional dealing with pathogens already justifies entirely new therapeutic approaches. Surely soon a complete overhaul will be witnessed.

Homo sapiens, already a star performer, like a tennis player or a top manager, is becoming increasingly lonely. At his apparent peak, he might soon be made aware that possession of power, without a concomitant care for the myriads of species that accompanied him in his long evolution, will rebound on him.

Fortunately, worldwide opinion is heading towards a consensus; already at the kindergarten, children are taught not to impede nature's design work.

LISTEN TO AN ECOLOGIST TRYING TO AVOID REDUNDANCY

Increasingly, the view is shared that the globe as a whole functions as a network and in this context it may be considered as an organism. Ecology implies a reciprocal relationship between things. If this is a global property, it cannot be assumed that the immune system will be an exception. As a consequence, subjects manipulate the immune environment which, on the other hand, manipulates them. This raises the question: can the environment integrate immune interaction through ecological means?

Current risk factors are, of course, many. A good example would be the global deforestation currently taking place. It also occurred in the late Middle Ages, coincident with what was known as the Black Death, which killed about one-quarter of the population of Europe; in some places three-quarters of the population were wiped out. This should act as a strong reminder for us to maintain a state of alertness.

Fortunately, global communication allows the alarm to be sounded simultaneously to billions of people, alerting them to the dangers of human numbers and consumption excesses. Everywhere there is evidence that the younger generation is starting to respond to the threat. It might not take much time until the predatory forces become outweighed, thereby improving the outlook for all.

Usually immune-ecological effects inside a congested collective may be too feeble to create a protective system. Epidemiologists are looking for clues such as disease incidence and characteristics under different circumstances, changes associated with drastic turns of life, etc. If the wealth of factors in a so-called milieu-soup combines all the influences regarding immune developments, we may perceive minute signals, far smaller than the ones thought now to cause immune manipulation. Somehow, as inside the intimate family circle, gestures may suffice to increase our understanding. Over time, so we speculate, a malleable public immune-homeostasis might evolve, preventing people from over- or under-reacting.

OUR BIOLOGICAL DUTY: DEFENSIVE STRATEGY REVISION

IMMUNE SPLIT

When we take into account the ephemeral period of antibiotic efficacy caused by bacterial adaptations, and the circumstances in which improvement overrides pharmacological effects, it would seem a good idea to find out the truth about our speculations.

Not only in therapeutics, but also in the chess-match of international politics, it has been acknowledged that complete victory, that is to say, a total extermination of the aggressor, is not advisable for defense—in any case, it has almost never been achieved. In a

FIGURE 30 A MERE SUGGESTION: TO SPRAY MICROFLORA INTO AIRCRAFT CABINS AND HOSPITALS

hypothetical case, one would expect that after such a total victory another enemy would soon come to replace the former. So, in infectology we are trying to learn how to live together with the aggressor, carefully avoiding harm. Strategy should always be based on regulation. Thanks to the achievements of molecular biology, there is now the possibility of removing aggressive use of antibiotics, omitting lethal action and sparing the associated flora.

Since time immemorial immune systems have developed a well-honed perception not only for invaders but for manifold signals as well. Immunity's reliance on cognitive properties should be emulated by therapeutics as far as is possible. Detailed knowledge of the ways in which the immune response priority selection takes place is important. The accepted ideal is to attempt to eliminate danger without affecting anything else, thereby acting in a regulatory capacity.

Globalization causes the numbers of pathogenic germ species we deal with to increase. As acquired refractoriness to antibiotics and drugs are more than mere hints, there are many problems to be solved. On the other hand, suitable treatments for new threats may emerge. We must uncover all the means by which microbes disseminate multidrug resistance. Furthermore, complete information of predatory activities and how invaders coordinate and multiply are needed.

The preceding chapters have attempted to show that there is no such a thing as an insurmountable barrier between individuals. Increasingly, therapeutics takes advantage of the dynamics of drug transit through the skin—which once was considered as an inviolable barrier. Somehow, propitious circumstances may favor the evolution of mutually adapted immune activities.

To make our point, we need to consider the extremes. Crowded humanity implies a constant, relatively large amount of substance exchanges, thereby promoting a common immune behavior. On that basis it seems worthwhile exploring possible adaptations to new risks. The fact that it has not been done yet suggests that under life conditions that we feel as normal, a significance for community in the immune defense system is negligible. We firmly believe that in historical crises that was not the case. There is a suspicion that in many places the same thing happens even today. It seems clear that there is a dichotomy in the immune defense concept between what are considered to be appropriate living conditions and what are considered the worst. It is in the latter that we anticipate the influence of common immune behavior. Health policy should be conducted with an awareness of the two faces of immunity. Trying to look in depth for rational use, we would not like to miss any niche that is, presumably, receptive to a therapeutic influence by incoming molecular flow.

FIGURE 31 THE TWO FACES OF IMMUNITY

<u>IMMUNE SPLIT</u>

ISOLATED PEOPLE: IMMUNITY AS USUAL

WHEN HABITAT IS BECOMING CROWDED
influencing the immune structure ?
contributing to the self image ?

**** HOW TO ASSESS SOMATIC EXPERIENCE?**

**** *VIA PSYCHE***

**** *VIA STRESS***

**** *MILIEU-SOUP* SUSCEPTIBLE TO CHANGE**

CHANGING

IMMUNE BIAS ?
NONSPECIFIC IMMUNITY
IMMUNE COGNITIVITY
IMMUNE HOMEOSTASIS
POPULATION BIOLOGY

* DEFYING INFECTIOUS CHALLENGES ?
* MULTIDRUG RESISTANCES SETBACK
* ANTIGEN REPERTOIRE SHARING
* RESISTANCE SPECTRUM WIDENS (AGE GROUPS)
* MAINTAINED IMMUNE STIMULATION FEEDS ALERTNESS
*HOST–FLORA–PARASITE INTERACTIONS
* DECOY EFFECT ON PATHOGENS

FIGURE 32 IMMUNE SPLIT

FIGURE 33 CONNECTIONS

CONCLUSION

Aiming to Prevent Catastrophe

A SOUND RELIGIOUS FEELING HAS BEEN GOOD FOR *HOMO SAPIENS* SINCE TIME IMMEMORIAL. BUT WE MUST BEWARE OF THE IMMEASURABLE SUFFERING THAT IS CAUSED BY ITS FANATIC VARIANT.

Immune-entangling came into the limelight because of its deterministic chaos behavior, which includes modern criteria about unforeseeable changes dependent on initial conditions, often at thresholds far below those traditionally thought to obtain. As far as immunity is concerned, a lot of essential elements remain elusive, although, surprisingly, that does not appear to be very obvious in everyday routine in which a relative predictability prevails. We cannot say that there is much difference between immunologists and colleagues dedicated to dealing with, for example, common wounds. Extreme situations rarely develop, although 'bombshells' may come at any time. The subject is topical, and waiting for a thorough investigation. There is no need for any conflict between what has already been accepted for an individual's immune system network and what concepts chaos theory may subsequently suggest.

The other main issue concerns plague. To prevent catastrophe, carelessness must be replaced by huge investment. That antibiotics do not provide security is another matter. Current research on immune repercussions in confinement, be it in support of space flights or stress-conditioned behaviors, mainly concentrates on mind as a carrier. In such situations messages reverberate through the recipients' physiological systems. But we do not believe that immune interdependence should exclusively be attributed to an indirect *modus operandi* that belongs to the sphere of psychoneuroendoimmunology or an expanded variant of it. Other ways of connecting may be feasible as well. Screening the available information from millennia of what is now thought of as inhumane living conditions—whether from antiquity or from today's shanty towns—would possibly bring to light key survival factors.

As has been discussed at length, there is no such thing as an insurmountable barrier between subjects. Polluted milieus in packed habitats pool virtually unlimited antigenic information, often complicated through associations or puzzling camouflage. As part of evolution, many propitious factors gradually promote immunity, among which importance is given to the fact that all species identify an individual self. To be able to thrive, all subjects should be able to identify what belongs to them and what does not. But here we postulate that self- or non-self targeting does not necessarily derive exclusively from inside conventional body limits. Nature seldom allows individuals to be entirely isolated. Small compacted communities ('subunits') are widespread in biology and deserve special attention in the defense context that immunology comprises. They perhaps played a role in survival. There is nothing new in the idea that an individual's intrinsic immune system cannot remain entirely independent.

Focusing on immunity *per se*, in the body, evolution has created a dispersed network that departs from innate immunity. In Man, the immune system achieved a fine discrimination between huge numbers of signals, hostile intruders among them. For the time being we are not aware if this comprehensive system assimilates xenogenic messengers as supportive or detrimental. The question arises whether proximity allows individuals to develop some kind of common immune behavior by integrating the whole set of factors—both those already discussed and those yet to be envisaged.

A clear distinction between living without restrictions on space and in confinement has been made. We know that children change their behavior when they go to school, as do people when they join the armed forces. Once an individual is locked inside an overcrowded prison, immune performance should also undergo an adaptation process. As often happens in biology, amid disruption stemming from the collapse of normality, compensatory mechanisms may emerge. It might be wrong to generalize, but in terms of evolution, collective handling appears to be advantageous. Even microbes seek protection, exchanging isolation for integration within a colony. Our immune defenses might share a similar process.

Without rejecting the placebo effect outright, we consider the expansion of the immune network, by environment sharing, together with the convergence of flora from the oral, intestinal, genitourinary and skin habitats, plus many organisms that the particular pollution attracted, as factors of potential relevance to the influence of the immune system. As the molecules from one's fellows become incorporated with one's own, interactions with cells or substances in the host may be of functional significance.

Finally, a word about environments. A 'milieu-soup' comprises, in addition to the well-known peptides, immunoglobulins, soluble receptor molecules, etc. relating to the immune network, a range of ancillary—apparently unrelated—molecules such as hormones, neuropeptides and conjugated apoproteins, all of which should be carefully evaluated. By habitat trafficking, we not only have in mind molecules and micro-organisms, but also cells that the hosts might release. It may not be far-fetched to consider that, potentially, all the elements together form another, minor immune medium alongside tissues and body fluids. Modern physics provides the possibility of new insights into such a system.

At the prospect of a post-antibiotic era even the faintest indications for a positive counterbalance should be assessed for potential expansion leading to therapeutic benefit. A good start might be to confront the extremes of the Gaussian distribution curve in terms of the conditions in which the communities live.

Epidemiological dynamics, expansive or reverting, are of interest. As we have already seen, the invaders' dissemination as well as the host's immune behavior are strongly influenced by environmental changes. Many sources point to the necessity to redefine the role of community in patterns of disease transmission and dissemination.

The characteristic flare-ups and -downs in the course of autoimmune disease have been discussed, and these point to fluctuations in the individual's immune profile, with the possibility of similar effects—whether beneficial or adverse—accruing from overcrowding in the host's surroundings.

Autoimmune manifestations are often associated with AIDS. Red-cell-associated auto-antibodies have been found especially frequently in terminally ill AIDS patients from the Muñiz Hospital that takes care of sufferers from the disease in Buenos Aires. González and his group worked day and night to provide invaluable information about the frequency of anti-erythrocytic antibodies at three levels of patient-crowding, one of them in confinement. The sad record of the hospital and, even worse, of the inmates is that in the two months covered by the study all the patients died of terminal AIDS. Overwhelmed by 150 000 annual admissions and huge mortality, and lacking minimal hospital structure (in one short period the blood-bank shut down twice), treatment often was not available or, at least, not duly recorded, nor were the associated diseases (for example, the ever-present tuberculosis). The rates speak for themselves. Any realistic approach to the facts we deal with cannot deny social connotations. To conduct serious research in the social holocausts, of which, regrettably, there are still many such around the world, was not possible. Nevertheless, the information obtained through this audacious attempt should be taken seriously enough to continue the project.

Microchimerism has become intensely scrutinized; but it needs to be checked whether close contacts are, perhaps, not restricted to molecular or bacterial particle interactions among hosts. Transplantation made chimerism a fact of life and increasingly we learn about microchimerism induced by simple blood transfusions. What has to be discovered is whether cell exchange also has a chance to occur under conditions of close confinement; at the moment we would like to investigate subjects with clear-cut immune deficiency (the immature, aged or diseased).

It seems important precisely to locate the individual in the immune context. It often happens that important new insights temporarily become overrated and tend to block other proposals whose importance might not necessarily have been less. For example, if immune interdependence should become a recognized fact, perhaps some role will come to light for those with whom we interact, flora and other bacteria, pets, etc., perhaps as a complement of the current self-centered exclusivity.

So far, adaptive immunological implications for overcrowding have not been credited. Immunologists' success stems from dealing with the internal network context, but the allusions brought forward here suggest that immunity is not necessarily that restricted. Even if available information is fragmentary in the extreme, we envisage that overcrowding in an immune–ecological niche offers the potential for a spectrum of consequences for good or for bad.

Perhaps we will now see a change in Man's predatoriness, although it could well take a considerable amount of time. It is, however, exciting to witness the enthusiasm with which young people assume responsibility, very far from the classical assumption that they are only either innocent or cruel.

References

1. Becerro MA, Turon X, Uriz MJ. Natural variation of toxicity in encrusting sponge *Crambe crambe* (Schmidt) in relation to size and environment. *J Chem Ecol* 1995; 21: 1931
2. Garrett L. *The coming plague*. New York: Penguin Books, 1994
3. Grady D. Quick-change pathogens gain an evolutionary edge. *Science* 1996; 274: 1081
4. Cohen IR. The cognitive paradigm and the immunological homunculus. *Immunol Today*. 1992; 13: 490–4
5. Ames BN, Profet M, Gold LS. Dietary pesticides (99.99% all natural). *Proc Natl Acad Sci USA* 1990; 87: 7777–81
6. de Boroviczény CG (ed). Standardization in Haematology. *Bibl haemat* 1966; 24: 186–93
7. Nowak MA, Bangham CRM. Population dynamics of immune responses to persistent viruses. *Science* 1996; 272: 74–9
8. Dennett DC. Darwin's dangerous idea. *The Sciences* 1995; 35: 34–40
9. Brack M. Metal clusters and magic numbers. *Sci Amer*, 1997; 31–35
10. Gleick J. *Chaos*. London: Abacus, 1988
11. *Encyclopedia Brittanica*. London, 1992; 25: 562–651
12. Gaardner J. *Sophie's world*. London: Phoenix House 1995
13. McKeon R. *The philosophy of Spinoza: the unity of his thought*. Woodbridge, Conn.: Ox Bow Press, 1987
14. Levin DM, Solomon GF. The discursive formation of the body in the history of medicine. *J Med Philos* 1990; 15: 515–37
15. Scapagnini U. Psychoneuroendocrinoimmunology: the basis for a novel therapeutic approach in aging. *Psychoneuroendocrinology* 1992; 17: 411–20
16. Levin AS, Byer VS. Environmental illness: a disorder of immune regulation. *State Art Rev Occup Med* 1987; 2: 669–81
17. Davies H. *Introductory immunology*. London: Chapman & Hall, 1997; 26–9
18. Adcock FE. *Thucydides and his history*. Cambridge: Cambridge University Press, 1963
19. Jerne NK. Towards a network theory of the immune system. *Ann Immunol* 1974; 125c: 373–89
20. Elder ME, Lin D, Clever J *et al.* Human severe combined immunodeficiency due to a defect in ZAP-70 a T cell tyrosine kinase. *Science* 1994; 264: 1596–9
21. Blalock JE, Harbour-McMenamin D, Smith EM. Peptide hormones shared by the neuroendocrine and immunologic systems. *J Immunol* 1985; 135(2 suppl): 858s–861s
22. Muraille E, Thieffry D, Leo O *et al.* Toxicity and neuroendocrine regulation of the immune response: a model analysis. *J Theor Biol* 1996; 183: 285–305
23. Benjamini E, Leskowitz S. *Immunology*, 2nd Edition. New York: Wiley-Liss, 1991: 302–9
24. Davis H. *Introductory immunobiology*. London: Chapman & Hall, 1997: 258–70
25. Weiss A, Littman DR. Signal transduction by lymphocyte antigen receptors. *Cell* 1994; 76: 263–74

26. Rowland-Jones S, Sutton J, Ariyoshi K *et al.* HIV-specific cytotoxic T-cells in HIV-exposed but uninfected Gambian women. *Nature Medicine* 1995; 1: 59–64
27. Hofstadter D. Cited by *The Guardian* [London] 1997; 8 August
28. Colovai AI, Molajoni, ER, Cortesini R, *et al.* New approaches to specific immunomodulation in transplantation. *Int Rev Immunol* 1996; 13: 161–72
29. Coutinho A. The network theory: 21 years later. *Scand J Immunol* 1995; 42: 3–8
30. Dixon RA, Lamb CJ. Molecular communication in interactions between plants and microbial pathogens. *Annu Rev Plant Physiol Plant Mol Biol* 1990; 41: 339–67
31. Sutherland S. *Irrationality: the enemy within.* London: Constable and Co., 1992
32. Cohen IR. The cognitive paradigm and the immunological homunculus. *Immunol Today* 1992; 13: 490–4
33. Ader R, Cohen AR, Felten D. Psychoneuroimmunology: interactions between the nervous system and the immune system. *Lancet* 1995; 345: 99–103
34. Teshima H, Sogawa H, Kihara H *et al.* Changes in populations of T-cell subsets due to stress. *Ann NY Acad Sci* 1987; 496: 459–66
35. Jessop JJ, Gale K, Bayer BM. Enhancement of rat lymphocyte proliferation after prolonged exposure to stress. *J Neuroimmunol* 1987; 16: 261–71
36. Cohen S, Tyrrell DA, Smith AP. Psychological stress and susceptibility to the common cold. *N Eng J Med* 1991; 325: 606–12
37. Berkman LF, Syme SL. Social networks, host resistance and mortality: a nine-year follow-up study of Alameda County residents. *Am J Epidemiol* 1979; 109: 186–204
38. Hennessy MB. Presence of companion moderates arousal of monkeys with restricted social experience. *Physiol Behav* 1984; 33: 693–8
39. Gust DA, Gordon TP, Brodie AR *et al.* Effect of companions in modulating stress associated with new group formation in juvenile rhesus macaques. *Phys Behav* 1996; 59: 941–5
40. Kingston SG, Hoffman-Goetz L. Effect of environmental enrichment and housing on immune system reactivity to acute exercise stress. *Physiol Behav* 1996; 60: 145–50
41. Riggs DS. *Control theory and physiological feedback mechanisms.* Baltimore: Williams & Wilkins, 1970
42. Flavius J. The Jewish war. In *The Great Histories.* New York: Washington Square Press, 1965: 77–336
43. Schmitt DA, Schaffer L. European isolation and confinement study. Confinement and immune function. *Adv Space Biol Med* 1993; 3: 229–35
44. Shark repellents. In *Sensory Reception.* McHenry, ed. *The New Encyclopaedia Britannica*, London, 1992; V27: 137
45. Pawlik JR. Marine invertebrate chemical defenses. *Chem Rev* 1993; 93: 1911–22
46. Maibeche-Coisne M, Sobrio F. Pheromone binding proteins of the moth *Mamestra brassicae*: specificity of ligand binding. *Insect Biochem & Mol Biol* 1997; 27: 213–21
47. Taylor LK, Wang HC, Erikson RL. Newly identified stress-responsive protein kinases, Krs-1 and Krs-2. *Proc Natl Acad Sci USA* 1996; 93: 10099–104
48. Baird RC, Johari H, Jumper GY. Numerical simulation of environment modulation of chemical signal structure and odor dispersal in the open ocean. *Chem Senses* 1996; 21: 121–34
49. Steinbrecht RA. Are odorant-binding proteins involved in odorant discrimination? *Chem Senses* 1996; 6: 719–27
50. Furlow FB. The Smell of Love. *Psychology Today.* March/April 1996: 38

51. Medzhitov R, Preston-Hulburt P, Janavay Jr CA. A human homologue of the *Drosophila Toll* protein signals activation of adaptive immunity. *Nature* 1997; 388: 394–7
52. Fearon DT. Seeking wisdom in innate immunity. *Nature* 1997; 388: 323–4
53. Good RA. Structure-function relations in the lymphoid system. In Bach FH, Good RA, eds. *Clinical immunobiology* V I. New York: Academic Press, 1972: 1–28
54. Christensen TA, Sorensen PW. Pheromones as tools for olfactory research. Introduction. *Chem Senses* 1996; 21: 241–3
55. Banerjee BD, Koner BC, Ray A. Immunotoxicity of pesticides: perspectives and trends. *Indian J Exp Biol.* 1996; 34: 723–33
56. Bach JF. Cytokine-based immunomodulation of autoimmune diseases: an overview. *Transplant Proc.* 1996; 28: 3023–5
57. Carr DJ, Rogers TJ, Weber RJ. The relevance of opioids and opioid receptors on immunocompetence and immune homeostasis. *Proc Soc Exp Biol Med* 1996; 213: 248–57
58. Davies H. *Introductory immunobiology*. London: Chapman & Hall, 1997: 271–7
59. Nydegger U. Video *Chasing after the molecule.* Wien: Good Day Film Production, Kräftner & Wagenhofer, 1995
60. Gallagher RB, Miller LJ. The elements of immunity [editorial] *Science* 1996; 272: 13
61. Cohen IR. Kadishman's tree, Escher's angels and the immunological homunculus. In Coutinho A, Kazatchkine MD, eds. *Autoimmunity, physiology and disease.* New York: Wiley-Liss, 1993
62. Schlitt HJ. Is microchimerism needed for allograft tolerance? *Transpl Proc* 1997; 29: 82–4
63. Stewart J, Varela FJ, Coutinho A. The relationship between connectivity and tolerance as revealed by computer simulation of the immune network: some lessons for an understanding of autoimmunity. *J Autoimmun* 1989; 2 Suppl: 15–23
64. Horák I. Interleukin-2–knockout mice: a new model to study autoimmunity and self-tolerance. Sb-Lek 1996; 97: 25–8
65. Theofilopoulos AN. The basis of autoimmunity. Part II. Genetic predisposition. *Immunol Today* 1995; 16: 150–9
66. Gross WL, Csernok E, Szymkowiak CH. Antineutrophil cytoplasmic autoantibodies with specificity for proteinase 3. In Peter JB, Shoenfeld Y, eds. *Autoantibodies.* Amsterdam: Elsevier, 1996: 61–7
67. Johnson HM, Torres BA, Soos JM. Superantigens: structure and relevance to human disease. *Proc Soc Exp Biol Med* 1996; 212: 99–109
68. Khamashta M, Petri M. Antiphospholipid antibodies hasten atheroma. [Editorial comment] *Lancet* 1996; 348: 1088
69. Lepennec PY, Lefrere JJ, Rouzaud AM *et al.* Red cell autoantibodies in asymptomatic HIV-infected subjects [letters]. *Transfusion* 1989; 29: 465–6
70. Gupta S, Licorish K. The Coombs' test and the acquired immunodeficiency syndrome. [Letter] *Ann Int Med* 1984; 100: 462
71. Human immunodeficiency virus (HIV) infection classification: MMWR 1–20 S 1987; 36S: 1–20S
72. Ewald P. *Evolution of infectious disease.* Oxford: Oxford University Press, 1994
73. Wills C. *Yellow fever—Black Goddess: the coevolution of plagues.* London: Addison-Wesley, 1996
74. Cherkasskii BL. The epidemic process as a system. I. The structure of the epidemic process. [Epidemicheskii Protsess Kak Sistema. Soobshchnie I. Structura] Zh *Mikrobiol Epidemiol Immunobiol* 1985; (3): 45–51

75. Benjamini E, Leskowitz S. *Immunology*, 2nd Edition. New York: Wiley-Liss, 1991: 302–9
76. Jack DB. Getting ready for the next influenza pandemic. [Editorial comment] *Lancet* 1996; 347: 1252
77. Featherstone C. Is *Escherichia coli* 0157: a superbug of mere sensation? *Lancet* 1997; 349: 930
78. Lassiter HA, Robinson TW, Brown MS *et al.* Effect of intravenous immunoglobulin G on the deposition of immunoglobulin G and C3 onto type III group B streptococcus and *Escherichia coli* K1. *J Perinatol* 1996; 16: 346–51
79. Galimand M, Guiyoule A, Gerbaud G *et al.* Multidrug resistance in *Yersinia pestis* mediated by transferable plasmid. *New Eng J Med* 1997; 337: 677–80

Previous Publications by the Author

Azcarruns CM, Hurtado L, Rewald E. Red cell transfusion in infected infants at altitude. Experimental basis IV for 'substitutive-inhibitory transfusions'. *Helv Paediatr Acta* 1969; 24: 420–2

Rewald E, Suringar F. 'Substitutive-inhibitory gamma globulin therapy' as prevention of stillbirth in Rh-incompatability. *Acta Haematol* 1965; 34: 209–14

Rewald E, Suringar F. Tolerance of intravenous gamma globulin. *Bibl Haematol* 1968; 29: 337–9

Rewald E. Intravenous gammaglobulin to prevent stillbirth in Rhesus incompatibility [Letter]. *Lancet* 1985; ii: 208

Rewald E, Morell A, eds. *Immunomodulation by intravenous immunoglobulin.* Casterton: Parthenon Publishing, 1993

Rewald E. IV IgG in the treatment of D-alloimmunized pregnant women. [Historians' Corner]. *Transfusion Today* 1991; 11: 8–9

Rewald E. A vascular effect of IgG therapy may prevent transplacental red cell leakage and spontaneous thrombocytopenic haemorrhages. *Med Hypotheses* 1990; 33: 193–5

Rewald E, Pizzolato M, Bragantini G, Saito N. Plateless-bleeding prevention by intravenous IgG (IVIG) until IgG complexes with IgM. *Int J Hematol* 1996; 64: s53

Saito N, Rewald E, Takemori N *et al.* Ultrastructural localization of human IgG in bone marrow cells after intraperitoneal injection of human IgG in mice; comparison with human albumin. *Int J Hematol* 1996; 64: s118

Rewald E, Saito N, Kohgo Y. Spontaneous bleeding prevention by gammaglobulin infusions? XVI Congr. Intern. Hemostasis y Trombosis, *Rev Iberoamer Tromb & Hemost* 1995; 8: (suppl. 3): 101–2

Rewald E. Repeated low doses of anti-Rhesus gammaglobulin in aplastic anaemia [Letter] *Lancet* 1987; ii: 795

Rewald E. Are there options for donor-derived Anti-D Preparations other than to prevent (Rh)D sensitization? *Transfusion Sci* 1995; 16: 383–9

Rewald E, González C. Combining IgG and interferon alpha-2 in chronic hepatitis C virus infection. *Transfusion Sci* 1996; 17: 433–6

Rewald E, Francishetti MM. Combining Interferon Alpha-2 with High-Dose IgG, IgM, and IgA in Myelofibrosis with Trisomy 1. [Letter] *Am J Hematol* 1997; 54: 340

Index